The 21st Century

Countdown To Eternity

by

Jim Raposa

authorHOUSE™

1663 LIBERTY DRIVE, SUITE 200
BLOOMINGTON, INDIANA 47403
(800) 839-8640
WWW.AUTHORHOUSE.COM

First published by AuthorHouse 09/06/05

ISBN: 1-4208-6374-6 (sc)

Printed in the United States of America
Bloomington, Indiana

This book is printed on acid-free paper.

Dedications

To my friend and mentor Dr. Lowell Davey, who as a young man was willing to set aside an opportunity to become a highly successful businessman, and commit his life to our Lord and Savior in full-time Christian service. It was his teaching and inspiration that led me to begin an extended study of Bible prophecy, which has led to the writing of this book.

Also, to Dr. Keith H. Davey, whose vision for reaching and teaching men and women in the armed services, ignited in me the desire to study the things mentioned in these chapters, so that I could share them with you through the printed page.

In Memoriam

To Chaplain William R. "Bill" Griswold and his wonderful wife whom I came to know as "mom". Chaplain Griswold, after having served the Lord for nearly a lifetime of military service, chose to go the "extra mile" by opening a small Christian Servicemen's' Center near the base at Great Lakes, Ill. It was there that I was introduced to Jesus Christ as Lord and personal Savior.

TABLE OF CONTENTS & CHAPTER DESCRIPTIONS

This chapter contains Scripture and related discussion to verify that the Bible does indeed teach that Christ will return and that His first advent (which we now celebrate as Christmas) did not end His involvement with mankind.

This chapter discusses the conditions we see in the world today and compares them to the "Days of Noah" as foretold in the Gospels

Here we discuss the current state of apostasy in the organized church today and the lack of concern on the part of the Christian community for the real mission of the church.

In chapter 4 we examine the prophesy in Ezekiel 38 & 39 and discuss how it matches up to the alignment of the nations today.

"Seventy Weeks" looks in depth at the Seventy Week Prophecy in Daniel chapter 9 and it's relationship to our calendar today. We talk about how we are currently in a prophetical parenthesis in the Seventy Week Prophecy, how the sixty nine weeks already past were fulfilled exactly to the day, and how the yet future Seventieth Week is going to play out.

Chapter six starts out by directing the reader to Daniel 12:4 which predicts an explosion of knowledge and travel that has taken place only during the last century and mostly during my lifetime. (The last fifty years) In 100 years we have gone from horse drawn carriages to jets that fly faster than a horse can imagine, and we have come to a point where technology now in place makes the one-world system of government and financial management described in Revelation a very real possibility.

Chapter seven deals with an often-misunderstood passage of the Bible found in Revelation Chapter 13, which deals with the "Number of the Beast". How can anyone expect that everyone in the world would be willing to have a special mark in order to buy or sell anything? This chapter offers some suggestions about the technology already in place that would make that possible.

This discussion shows how we, who are a part of the Church, fit in to the overall prophetic picture. It explains how God has temporarily set His Chosen People, the Jews, aside for a season known as the "Times of the Gentiles" mentioned by Christ in Matthew 25. We will see that it began with the destruction of Jerusalem by Rome in 70 A.D. and how it will end sometime shortly after Israel returns to her homeland, an event that has begun *during my lifetime!*

In this chapter we extend our discussion of the technological explosion of the last century with the emphasis being on the time period in which this foretold explosion of knowledge has taken place. We have gained more insight into the Creator's plan and His methods in the last 100 years than in the preceding 5900 and acknowledge Him less! We know about atomic and molecular structure, genetic engineering, how the brain functions and far too much to even cover in this paragraph and even in the book. *All in my lifetime!*

Chapter ten talks about how there have been six periods of 1000 years since man's placement on this planet as recorded in Genesis. It shows how there has been a catastrophic event at every 2000-year point and how we are now at one of those points as we enter the seventh millennium. It also discusses how six is used throughout Scripture as the number of man and seven as God's number and considers the possibility that man's time may be drawing to a close.

Chapter 11 deals with statements made in Daniel and Revelation concerning Bible prophecy. In Daniel the prophet is told to seal his prophecy until the last days. In Revelation, John is told to not seal his prophecy. Why the difference? Is this a contradiction? This chapter deals with these issues and clearly explains them.

The Olivet Discourse is discussed in detail, *in context* (now that's different!) We can see the fact that we are indeed the generation that will see the things in that prophecy take place. We then must determine how long a generation lasts. We throw out the dictionary and discuss the Bible's definition of a generation showing how these events are yet ahead of us, but not by much.

In this chapter we put it all together, showing how the first 10 chapters all fit together like a nice puzzle and how chapter 11 ties it to our lifetime, probably in the next 20 years. It is important to understand however that the mission of this book is not to set a date for Christ's return, but rather to show that there is a greater than ever possibility that our generation will be the one that will meet the Upper-Taker instead of the Undertaker!

What if everything we have talked about in the preceding chapters is true and there is a possibility that Christ could return in my or your lifetime? What preparations should we make? This chapter takes the basic plan of salvation and in a very easy to understand and very palatable way presents it to the lost and challenges Christians to live in anticipation of the very real possibility that the return of Christ in the Rapture will be very soon... *Perhaps Today!*

"The 21st Century" ©

"Why I Believe Christ will Return in my Lifetime"

Preface

Every now and then, someone will write a book, teach a class, or preach a sermon that will supposedly tell us when Christ will return. Some will say it's going to be on a certain day, in a certain place, or even at a certain time. As we begin the third millennium since the Advent of Jesus Christ, Second Coming guessing has taken on a new fervency. Almost every day you hear of somebody who now "knows" when it will happen.

Please understand that this is neither the purpose nor intention of this book. Christ Himself warns us that it is not for us to know the day and hour of His return, but rather to keep busy about His work in anticipation of it. Some writers have even gone so far as to misinterpret that verse and say that He only told us we would not know the day and hour, but could know the year. Never read anything into Scripture that isn't already there. That's a sure way to generate a lot of unnecessary confusion.

You may ask then, "Why all the numbers?" which you will find as you read through the latter chapters of this book. I've put those in there so you can see that there are several reasons why I think *this generation* will be the one that sees the long anticipated Rapture which will usher in the Great Tribulation. It is more likely to happen *in our lifetime* than in any generation that has lived on earth since His first advent.

When? Perhaps today! Maybe it'll be sometime in the next few years as this book suggests. (Note, I said *suggests*, not states) It could also be in the next few minutes or even the next few seconds. It could be in the next few decades. So what if the next couple of decades pass and it still hasn't happened? We should always live and work as if it's going to be in the next few minutes. That's what we're told to do by the Author of Scripture and I've no good reason to do otherwise.

CHAPTER ONE:
IS HE REALLY COMING BACK?

During my childhood, I was brought to a church where nothing was ever mentioned about the return of Jesus Christ. We considered ourselves "Christians" because we believed in Jesus Christ, believed (intellectually) that He came to earth, being born of a virgin, was crucified, died, and was buried. We were taught about the Resurrection, Ascension, and His eternal presence at the right hand of the Father. We were even taught that we could one day join Him there if we lived right. (*That, of course, is not found in the Bible!*) But nothing was ever mentioned of a Second Coming of Jesus Christ.

The Bible was not a book for people to read and understand. That was the mission and duty of the Church. As a teenager, I met a Navy Chaplain who taught me differently. I began to see that the Bible was a special book written under Divine Inspiration by men of God for us to live by. And as I began to study that wonderful Book, I began to learn new things that were never taught to me before. One of the most exciting of these things, is that Christ would return, and not just once. He would come in the clouds to take those who love Him to Heaven.[1] Seven years later, He would return

---

[1]. *I Thessalonians 4:13-18* - And now, brothers and sisters, I want you to know what will happen to the Christians who have died so you will not be full of sorrow like people who have no hope. 14For since we believe that Jesus died and was raised to life again, we also believe that when Jesus comes, God will bring back with Jesus all the Christians who have died. 15I can tell you this directly from the Lord: We who are still living when the Lord returns will not rise to meet him ahead of those who are in their graves. 16For the Lord himself will come down from heaven with a commanding shout, with the call of the archangel, and with the trumpet call of God. First, all the Christians who have died will rise from their graves. 17Then, together with them, we who are still alive and remain on the earth will be caught up in the clouds to meet the Lord in the air and remain with him forever. 18So comfort and encourage each other with these words.

in triumph and glory to rule the world for 1000 years.[2] You will learn more about these great prophetical events yet to take place as you progress through this book.

The Old Testament and New Testament are replete with statements about the Second Coming of the Messiah. In the Old Testament they were not specifically referred to in that way, and the Hebrew scholars often made the mistake of looking for just one coming of a Messiah and never comprehended the possibility that He would come first as Messiah-Redeemer, a sacrifice to pay for our sins, then come again in glory as Messiah-King to bring an end to rule by sinful men and establish a kingdom here on earth. And therein lay the problem. The Hebrew people had a history of affliction under the hand of tyrannical rulers. First it was slavery in Egypt. God sent the deliverer, Moses. The Jews still commemorate that occasion today with the celebration of the Passover. Then came 70 years of captivity under the Babylonian Empire as recorded in the Book of Daniel. At the time Christ was born they were being ruled under the strong hand of Rome. Always under oppression, seemingly forsaken by the God who called them His own. And so, because of their oppression they were forever expecting fulfillment of God's promised Messiah. They had completely overlooked His promise of a Messiah-Redeemer and so when "He came unto His own, His own received Him not".[3]

The Bible's promises started as far back as the Garden of Eden. God told Eve, after her unfortunate confrontation with Satan, that one would come who would bruise the Serpent's head and the Serpent would bruise His heel, a clear prophecy to the atoning work of Christ.[4]

Later in the book of Genesis, Abraham was told to take his only son, Isaac, and offer him as a sacrifice to God, but was spared that agony at the last minute partly as a test of his obedience, but more so as a type of the sacrifice that God would require of His own Son on Calvary.[5] In Isaiah 53

[2.] ***Revelation 20:4*** - Then I saw thrones, and the people sitting on them had been given the authority to judge. And I saw the souls of those who had been beheaded for their testimony about Jesus, for proclaiming the word of God. And I saw the souls of those who had not worshiped the beast or his statue, nor accepted his mark on their forehead or their hands. They came to life again, and they reigned with Christ for a thousand years.

[3.] ***John 1:11*** - Even in his own land and among his own people, he was not accepted.

[4.] ***Genesis 3:15*** - From now on, you and the woman will be enemies, and your offspring and her offspring will be enemies. He will crush your head, and you will strike his heel."

[5.] ***Genesis 22*** - 1Later on God tested Abraham's faith and obedience. "Abraham!" God called. "Yes," he replied. "Here I am." 2"Take your son, your only son--yes, Isaac, whom you love so much--and go to the land of Moriah. Sacrifice him there as a burnt offering on one of the mountains, which I will point out to you." 3The next morning Abraham got up early. He saddled his donkey and took two of his servants with him, along with his son Isaac. Then he chopped wood to build a fire for a burnt offering and set out for the we have a detailed account of how Christ would be sacrificed for the sins of the world. And of course, that is what the New Testament is primarily about.

But there are numerous references to the coming of Christ as Messiah-King throughout the Old and New Testaments. Christ Himself spoke of it in every one of the four gospels. In Matthew it is discussed in more detail than any other of the four gospels. Matthew 24 and 25 give a lot of detail about it and an entire chapter in this book covers that teaching. In Mark Christ discusses it in chapter 13; in Luke it's chapter 17, and in John it's chapter 14, where we find one of the clearest passages that teach that Christ will return for His own one day. As we study on we find it again in Acts, I & II Thessalonians, II Peter, and of course, Revelation. Many of these passages will be noted as we continue.

There is no doubt as to His first coming and it's purpose. And as you read this book and study the Scriptures, you should be equally convinced that He will come again. When? Read on....

---

place where God had told him to go. 4On the third day of the journey, Abraham saw the place in the distance. 5"Stay here with the donkey," Abraham told the young men. "The boy and I will travel a little farther. We will worship there, and then we will come right back." 6Abraham placed the wood for the burnt offering on Isaac's shoulders, while he himself carried the knife and the fire. As the two of them went on together, 7Isaac said, "Father?" "Yes, my son," Abraham replied. "We have the wood and the fire," said the boy, "but where is the lamb for the sacrifice?" 8"God will provide a lamb, my son," Abraham answered. And they both went on together. 9When they arrived at the place where God had told Abraham to go, he built an altar and placed the wood on it. Then he tied Isaac up and laid him on the altar over the wood. 10And Abraham took the knife and lifted it up to kill his son as a sacrifice to the LORD. 11At that moment the angel of the LORD shouted to him from heaven, "Abraham! Abraham!" "Yes," he answered. "I'm listening." 12"Lay down the knife," the angel said. "Do not hurt the boy in any way, for now I know that you truly fear God. You have not withheld even your beloved son from me." 13Then Abraham looked up and saw a ram caught by its horns in a bush. So he took the ram and sacrificed it as a burnt offering on the altar in place of his son.

CHAPTER TWO:
THE CONDITION OF THE WORLD

When I started elementary school in 1951, the most common discipline problems were things like chewing gum in class, not throwing waste paper in the trash cans, inattention to class work, running in the halls, being late for classes, and absence. Today, just over 50 years later, conditions are drastically different. Many school administrators now have to deal with things like illegal drugs, handguns, graffiti, profanity, rape, and student-teacher violence. Uniformed security guards patrol the halls of our public high schools and junior high schools, something unheard of when I attended high school. It makes one wonder where our society is heading. What has happened to our young people and why? Is this just another social cycle in America, or have we begun to degenerate into an animal like behavior where only the strongest of the species survives and lives to prey on the weaker ones?

The Bible tells us "There is no new thing under the sun". [1] Has this type of conduct among man ever occurred before? A careful examination of Scripture would indicate that it has, but that anytime there was such a sudden increase in evil living, it was always linked to a catastrophic judgment that would follow almost immediately. What about Sodom and Gomorra? God appeared to Abraham and told him of His plan to destroy Sodom & Gomorra. Abraham begged God's mercy and told him that he would spare the cities if ten righteous people could be found: only ten out of thousands! [2] However, there were not five; only Lot, his wife, and their daughters. Homosexuality was rampant. To publicly acknowledge

---

[1]. *Ecclesiastes 1:9* - History merely repeats itself. It has all been done before. Nothing under the sun is truly new.

[2]. *Genesis 18:32* - Finally, Abraham said, "Lord, please do not get angry; I will speak but once more! Suppose only ten are found there?" And the LORD said, "Then, for the sake of the ten, I will not destroy it." 33The LORD went on his way when he had finished his conversation with Abraham, and Abraham returned to his tent.

that one was homosexual was not considered a sin, but an "alternative lifestyle". Sound familiar? You know the rest of the story. Today we are still digging up the ruins of those cities and verifying that they were destroyed exactly as God said they would be.

What about the Tower of Babel? Located in the heart of what we know as Iraq today, it was a place where all the people of the then known world had gathered to build a great tower to reach the heavens. It was a place of idol worship, prostitution, heathen ritual and even human sacrifice. God looked down and saw the work of the children of men and destroyed the tower and confounded the languages of men so that they were scattered into the many tribes and nations as we now have today.[3] And what about Ninevah? Jonah reluctantly warned the Ninevites of the coming judgment and they repented, only to turn back to their wicked ways a few years later leading to the ultimate destruction of the city. We now know the destruction of Jerusalem in 70 AD by the Roman army to be a historical event, which happened exactly as Christ said it would.

And while were talking about judgments and sinful conduct, let's not forget to look at the Great Flood of Genesis Chapter 8 in which only Noah and his family were spared out of the entire population of the world. Men without God are constantly finding new ways to plunge into the depths of depravity. God's law, (the Ten Commandments) has little or no meaning to people when they choose to live in open rebellion against Him. "But as the days of Noah were, so also shall the coming of the Son of Man be."[4] (These are the words of Jesus Christ from Matthew 24).

Today, our public schools and colleges are teaching our young people that they came from monkeys. It stands to reason: if we teach them that they descended from animals, why shouldn't we expect them to act like animals? I wrote a small poem that goes something like this:

> Once there was a monkey
> Sitting in a tree
> Looked a little like you
> And a little like me

---

[3]. **Genesis 11:1,7-9** - 1At one time the whole world spoke a single language and used the same words; 7Come, let's go down and give them different languages. Then they won't be able to understand each other." 8In that way, the LORD scattered them all over the earth; and that ended the building of the city. 9That is why the city was called Babel, because it was there that the LORD confused the people by giving them many languages, thus scattering them across the earth.

[4]. **Matthew 24:37–39** 37"When the Son of Man returns, it will be like it was in Noah's day. 38In those days before the Flood, the people were enjoying banquets and parties and weddings right up to the time Noah entered his boat. 39People didn't realize what was going to happen until the Flood came and swept them all away. That is the way it will be when the Son of Man comes.

> He slipped on some sap
>> And fell into the Gap
> Broke off his tail
>> And let out a wail
>
> Put on a suit
>> And a tie to boot
> Wrote a book on Evolution
>> And called it the Solution
>
> The world believed
>> This stupid fool
> And now he's teaching
>> In a public school

In all fairness to our schools, I should mention here that some schools are now trying to offer alternative concepts as theories of the beginnings of civilization, such as "intelligent design" or "Theistic Evolution', in which evolution is still mentioned, but with the possibility that there is an infinite Creator out there who started it all. Parents are trying to require the schools to teach these theories on an equal basis with evolution, rather than treating any one of them as fact, as has been the case with evolution for many years now, After all, have you ever met anyone that was there when it happened? Can anyone you know absolutely prove, by <u>eyewitness</u>, that evolution is how the world started? I don't think so.

Morals have gone out the door and been replaced with what is "politically correct". Forty years ago it was a shame to be considered a homosexual. We had names for them when I was a child, which I would prefer to leave out of this text. Back then the word "gay" meant happy and full of joy. Now, in the last three decades of the 20th century we have redefined the word. "Gay" now means homosexual or lesbian. Instead of sin, it's an "alternative lifestyle". I have often heard it said that if God spares America, He would need to apologize to Sodom and Gomorra! Thirty years ago with a decision by the Supreme Court in the United States known as Roe vs. Wade, we began destroying our future work force. Since then over 30 million babies have been aborted. Today as you drive through most cities in America you see "Now Hiring" signs everywhere. An employer who placed an ad in the "help wanted" section of the paper ten years ago could expect to receive 50 to 100 applications or more, depending on the job. Today they are lucky to find a half dozen. We are facing an employment crisis in this country and it will only get worse as we continue to eliminate the potential workforce of the next generation. We have already begun to reap what we have sown. Abortion has been responsible for the death of more unborn babies than all the people murdered by Adolph Hitler in the Holocaust. In China, families having more than one child are subject to mandatory abortion. Mothers killing their own babies; something unheard of before this century, except in instances of human sacrifice which brought swift and certain judgment as already noted in this chapter.

Teen pregnancy, wife swapping, short-term marriages, divorces, adultery, and the like are all common occurrences. According to our Supreme Court, it's perfectly fine for unmarried people to have sexual relationships, as long as it is between "consenting adults". I can remember the fifties, when hotels and motels were monitored for the presence of illicit relationships and those doing so were criminally charged. While I will not comment any further on those days, I can look back and see that the divorce rate was lower and domestic violence was not as big a problem as it is today. As one well-known cigarette manufacturer advertises, "You've come a long way, baby!" They just fail to tell us in which direction. And if you think America has degenerated, just look at Europe. And now even the former iron curtain countries are plagued with social problems and crime.

The 20th century has been one of war. Since the year 1900, there have been more wars and mass destruction of human life than all the years of recorded time prior. For at least two major periods of time, the entire world was in global conflict, known as World War I and, the war to end all wars, World War II. (There have been at least four major wars since then!) The last world war ended with the unleashing of weapons of such destructive power that the entire earth could easily be destroyed with just a few major strikes. And not only have the major powers of the earth built up arsenals of such weapons, but there is even talk of a new more destructive weapon, designed to destroy life without the massive environmental damage of the atomic bomb. It's called the Neutron Bomb. Could it be that some day, perhaps during the final battle at Armageddon, we could see that brought into use? Possibly that is what is referred to in Revelation where the Angel of Death is told to destroy one third of the earth's occupants without causing any damage to the trees or vegetation.[5] Of course, when God chooses to execute such a massive judgment, he doesn't need our help. It's just interesting to notice that we have, only in this century, developed weapons capable of unleashing destruction on such a massive scale.

The Bible also mentions plagues that would result in loss of large segments of the earth's population.[6] Look at what AIDS is doing around

---

[5]. *Revelation 9:4* - They were told not to hurt the grass or plants or trees but to attack all the people who did not have the seal of God on their foreheads. 5They were told not to kill them but to torture them for five months with agony like the pain of scorpion stings. 6In those days people will seek death but will not find it. They will long to die, but death will flee away!

[6]. *Rev 6:1 – 9:21* (Note: Several different plagues are mentioned in this section of the Book of Revelation. Please look these up at your convenience, since they are to long to list here.)

the world today. A little virus so small that we can see it only with powerful microscopes, yet it can inflict suffering and destruction on people in numbers that are almost unbelievable. Is AIDS a judgment of God on homosexuality? You can decide that. It, like all plagues, takes a lot of innocent victims as it spreads. My point is not to argue it's judgmental aspects, but to show that when God chooses to deal with man He can do so without our help or intervention.

I believe that we are approaching, not just in America, but also on a world scale, the end of the last major cycle of sin and judgment on God's prophetic calendar. As we now begin the seventh millennium, we find ourselves in the days that Jesus spoke of where he said, "For as in the days that were before the flood, they were eating and drinking, marrying and giving in marriage, until the day that Noah entered into the ark, and knew not until the flood came, and took them all away; so shall also the coming of the Son of Man be." [7]

I have every reason to believe, from the study of Scripture that what we are experiencing now is the beginning of what will be the last major surge of sin and depravity that will occur before the Rapture, the great event in which those who know Christ as personal Savior and Lord will be removed and taken to Heaven.[8] The next major event after that will be what Bible students call the "Tribulation Period", [9] a period of seven years of the wrath of God against the sins of men, which will end with the visible return of Jesus Christ, who will judge the nations of the world.

---

[7]. *Matthew 24: 37-39* "When the Son of Man returns, it will be like it was in Noah's day. 38In those days before the Flood, the people were enjoying banquets and parties and weddings right up to the time Noah entered his boat. 39People didn't realize what was going to happen until the Flood came and swept them all away. That is the way it will be when the Son of Man comes.

[8]. *I Thessalonians 4:13-18* - And now, brothers and sisters, I want you to know what will happen to the Christians who have died so you will not be full of sorrow like people who have no hope. 14For since we believe that Jesus died and was raised to life again, we also believe that when Jesus comes, God will bring back with Jesus all the Christians who have died. 15I can tell you this directly from the Lord: We who are still living when the Lord returns will not rise to meet him ahead of those who are in their graves. 16For the Lord himself will come down from heaven with a commanding shout, with the call of the archangel, and with the trumpet call of God. First, all the Christians who have died will rise from their graves. 17Then, together with them, we who are still alive and remain on the earth will be caught up in the clouds to meet the Lord in the air and remain with him forever. 18So comfort and encourage each other with these words.

[9]. *Matthew 24:21-29* - For that will be a time of greater horror than anything the world has ever seen or will ever see again. 22In fact, unless that time of calamity is shortened, the entire human race will be destroyed. But it will be shortened for the sake of God's chosen ones. 23"Then if anyone tells you, `Look, here is the Messiah,' or `There he is,' don't pay any attention. 24For false messiahs and false prophets will rise up and perform great miraculous signs and wonders so as to deceive, if possible, even God's chosen ones. 25See, I have warned you. 26"So if someone tells you, `Look, the Messiah is out in the

Are you ready? We live in exciting times. But not only is the condition of the world and the moral landslide of this century a major factor, but as we will see in the next chapter, the condition of the Church is another great marker in God's record of time.

---

desert,' don't bother to go and look. Or, `Look, he is hiding here,' don't believe it! **27**For as the lightning lights up the entire sky, so it will be when the Son of Man comes. **28**Just as the gathering of vultures shows there is a carcass nearby, so these signs indicate that the end is near. **29**"Immediately after those horrible days end, the sun will be darkened, the moon will not give light, the stars will fall from the sky, and the powers of heaven will be shaken.

CHAPTER 3:
THE CONDITION OF THE CHURCH

When you read the Bible, you are also interpreting it to your own understanding of what it says. The best rule of Biblical Interpretation tells us that we should always take the Bible literally, in context, where possible and then look at any possible figurative application. In our study of these topics we will try to stay with that method. The Book of Revelation, however, presents several problems with literal interpretation. Is the Beast an animal? Or is it a man? Or is it a system? Are the plagues literal plagues? And what about the seven churches of Revelation chapters 1 to 3? Were they real churches? I believe that this is one passage where we should look at both the literal and figurative interpretation.

First of all, they were real churches. All were located in Asia Minor during the first century. But why would God leave us with a detailed history of seven churches in Asia Minor in a prophetical book like Revelation? Simply put, He wanted to tell us the story of the Church Age or the "Day of Grace"; the period of time from the Ascension of Christ to Heaven to the Rapture of the Church [1], the day in which the church will be removed from the earth and taken away to Heaven. This will then begin the dispensation known as the Tribulation Period[2], a time when sin and Satan are in total control and God's judgments are poured out upon the earth.

So, what has happened to the church as recorded in ancient history? How does it compare to the descriptions of the periods of progression in the Church Age given in Revelation 1 to 3? And where are we now? We can best answer these by looking at the descriptions of the churches listed in Revelation and briefly commenting on each of them.

[1.] *Ibid* – See Chapter 2, Note 8

[2.] *ibid* – See Chapter 2, Note 9

The first of these was the Church at Ephesus.[3] This was the church that had seen persecution and was tried and found faithful but had left its first love. This seems to point to the period from 20 to 200 AD, a time in which the Apostolic Age was drawing to a close. Following this we see the Church at Smyrna, which seems to point to the time of intense persecution by Rome that occurred from 200 to 316 AD. Martyrdom was at its peak and openly declaring one was a Christian meant death by anything from being beheaded to being thrown to the lions. After this we have the church at Pergamos, a church settled comfortably into the world in a church-state marriage that nearly led to its destruction. During the time of Constantine, a Christian emperor in Rome, the church enjoyed the benefit of governmental favor. But the great fall was yet to come. This covered a span of time from 316 to 500 AD. Some believe that this may be where Roman Catholicism had its roots, which ushered in a period of time in which the Roman Church became dominant, with only a small remnant of believers inside her walls. The conditions described in the Revelation Chapter 3 account of the Church at Thyatira seem to point to a period of time which spans 500 to 1500 AD, nearly a millennium, where there were no Bibles available to the common people in their own languages. The Roman Church was the sole authority in matters of faith and doctrine. After Constantine, pagan rulers took over the government at Rome, still the world power, and there introduced their heathen rituals and dogmas into the church's worship.

Perhaps it would be good to pause here for a minute and take a closer look at what happened and how it can and *will* happen again. During the time of Christ and during the first Century A.D., the most popular view of life was based on a mindset created by the philosophies of Plato. Plato was a Greek philosopher who lived from 428 BC to 348 BC. Plato's philosophy was the love of wisdom. This worldview held that anything spiritual was good and anything physical was bad. As the church spread abroad from Jerusalem to the Gentile (Greek & Roman) nations, the Gnosticism of the Greeks (pursuit of knowledge) and the political structure of the Romans (pursuit of power) began to influence the early church.

[3.] ***Revelation 2:1-7*** - "Write this letter to the angel of the church in Ephesus. This is the message from the one who holds the seven stars in his right hand, the one who walks among the seven gold lamp stands: **2**"I know all the things you do. I have seen your hard work and your patient endurance. I know you don't tolerate evil people. You have examined the claims of those who say they are apostles but are not. You have discovered they are liars. **3**You have patiently suffered for me without quitting. **4**But I have this complaint against you. You don't love each other or me as you did at first! **5**Look how far you have fallen from your first love! Turn back to me again and work as you did at first. If you don't, I will come and remove your lamp stand from its place among the churches. **6**But there is this about you that is good: You hate the deeds of the immoral Nicolaitans, just as I do. **7**"Anyone who is willing to hear should listen to the Spirit and understand what the Spirit is saying to the churches. Everyone who is victorious will eat from the tree of life in the paradise of God.

Some of the issues dealt with in the Pauline Epistles which are now part of the New Testament, show this was already happening as the New Testament was being penned. As the church was going thru its earliest stages, new ideas were being added, many of which resulted from the moral and social syncretism between society and the church.[4]

Here are just a few:

- The Incarnation: 2 John 7-11 deals with the incarnation of Christ, from which later emerged the worship and elevation of Mary.

- Resurrection: The philosophy of Plato didn't leave room for belief in a physical resurrection of the body. For this reason Paul wrote a very emphatic statement about it in I Corinthians 15, saying, "If there be no resurrection, our faith is in vain and ye are yet in your sins"

- Marriage: Physical relationships are bad according to the Plato mindset. From this emerged the concepts of clerical celibacy and monasticism, both widely practiced in the Roman Catholic Church today.

- Changed Authority: No longer "Scripture Alone" (Sola Scriptura) as the final authority for matters of faith and worship. Now the Catholic Church had become the sole authority and the declarations of its Papacy and "church fathers" became the new standard.

- Political Syncretism: The Roman government became the model for the church's system of government. From that concept came the creation of the Papacy, College of Cardinals, Bishops, and finally Priests and Deacons.

Can it happen today? Will it happen? After all with such a diverse spread of ideas and knowledge today, can it be repeated? The answer is yes. Without getting into a detailed study of the Book of Revelation, where we see a picture of a one-world religion with a person referred to as the "False Prophet" as its leader, I can tell you it's not only coming, but its roots are here already. The same social, moral, and political syncretism that nearly destroyed the true church and led the world into the "dark ages" is happening in our lifetime. Some churches are becoming centers of entertainment instead of places of worship. We all know that there are good church programs on TV that help communicate the Gospel message to those who are handicapped, sick or otherwise unable to attend. However, there are also church programs that amount to nothing more than entertainment. Music in the churches now spans the spectrum

---

[4] ***Dr. Mike Windsor***, CBI Class on Major World Religions 4/18/05.

from conservative to hard rock. Just pick the style of music you like, then find a church where that is the accepted style and you can have it all. Many churches in *all* denominations now offer "contemporary worship". Ecumenism tells us we all believe in God and if we also "believe" in Christ in some form or another, we can all worship and believe together, regardless of our denominational label. I expect that this type of syncretism where the worldview of society becomes the worldview of the church will lead to the apostate super church we see in Revelation 13.

From 1500 AD to 1700 AD we have a church striving to come out of idolatry into true worship in a time known as the reformation. This was the Church at Sardis.[5]

This time in history saw the invention of the printing press and the translation of the Bible from the Latin Vulgate into the language of the common people. Men like Luther, Wycliffe, Gulton, and many more paid dearly, some with their lives, to see the Gospel once again begin to spread across the then known world. Then came the great Revivals of the last two centuries. From 1700 to 1900 people were converted in large numbers. Men like Charles Wesley, D.L. Moody, Billy Sunday, and others traveled around the world with the Gospel message. But even with all this, an "open door"[6], Satan would have his professing church within the ranks of true Christians. Finally, comes the Church at Laodicea. The lukewarm church that was "rich and increased with goods, having need of nothing" Today many churches openly compete for the most beautiful building or the largest congregation, or the biggest television ministry, while missionaries are forced to live in poverty as they take the Word of God to the four corners of the earth. This appears to point to the period that started in the early 1900's and exists today. According to Scripture, there would come a great time of apostasy in which many churches would preach "another gospel"[7]. And although Matthew 24:4-12, taken in context, was not written

---

[5]. ***Revelation 3:1-6*** "Write this letter to the angel of the church in Sardis. This is the message from the one who has the sevenfold Spirit of God and the seven stars: "I know all the things you do, and that you have a reputation for being alive--but you are dead. 2Now wake up! Strengthen what little remains, for even what is left is at the point of death. Your deeds are far from right in the sight of God. 3Go back to what you heard and believed at first; hold to it firmly and turn to me again. Unless you do, I will come upon you suddenly, as unexpected as a thief. 4"Yet even in Sardis there are some who have not soiled their garments with evil deeds. They will walk with me in white, for they are worthy. 5All who are victorious will be clothed in white. I will never erase their names from the Book of Life, but I will announce before my Father and his angels that they are mine. 6Anyone who is willing to hear should listen to the Spirit and understand what the Spirit is saying to the churches.

[6]. ***Revelation 3:8*** - "I know all the things you do, and I have opened a door for you that no one can shut. You have little strength, yet you obeyed my word and did not deny me."

[7]. ***I Thessalonians 2:3*** - So you can see that we were not preaching with any deceit or impure purposes or trickery.

to the Church, it sure sounds like the times we live in today, in which "iniquity shall abound" and "the love of many shall wax cold". It seems like every time we pick up the paper we read about another church leader that has fallen into sin and resigned from his chosen place of ministry.

The study of these churches and what God said to each of them shows good and bad in each one. Even the lukewarm Church of Laodicea, the one that nauseated the Lord, had, and still has, the opportunity to abandon its worldly wealth and buy of God "gold tried in the fire" that it may still hold forth the Word of Life. There are still a few churches doing that today, but as the Church Age draws to a close, their numbers are decreasing every year.

In closing this chapter, I would like to point out three things. First of all, as we now find ourselves into the early 21ˢᵗ century, we see a technological explosion (discussed in a later chapter of this book) that will allow the Gospel to be preached worldwide.[8] Secondly, we should not overlook the fact that the great apostasy predicted in the Bible has come full circle.[9] The powerless church described in God's Word as "Having a form of Godliness, but denying the power thereof"[10] is in place as I write these words. Churches, with all their money and means, are doing little to reach the lost. Most Christians never share their "great secret" with anyone, much less lead a soul to Christ. And finally, getting back to our text in Revelation, there are no more church-history periods following Laodicea. The next thing on God's calendar is the trumpet.[11] Excited? You should be!

[8.] *Matthew 24:14* "And the Good News about the Kingdom will be preached throughout the whole world, so that all nations will hear it; and then, finally, the end will come."

[9.] *2 Timothy 4:3* For a time is coming when people will no longer listen to right teaching. They will follow their own desires and will look for teachers who will tell them whatever they want to hear.

[10.] *2 Timothy 3:5* They will act as if they are religious, but they will reject the power that could make them godly.

[11.] *Revelation 4:1* - Then as I looked, I saw a door standing open in heaven, and the same voice I had heard before spoke to me with the sound of a mighty trumpet blast. The voice said, "Come up here, and I will show you what must happen after these things."

CHAPTER 4:
THE ALIGNMENT OF THE NATIONS

Does the Bible have anything to say about the world's governmental structure in the last days? If so, what does it say? And where do we find it? For the answer to this question we will need a map, and a very important chapter in Bible prophecy, Ezekiel Chapter 38. In Ezekiel 37 God started out this prophecy by causing Ezekiel to have a vision of a valley full of dry bones. As Ezekiel watched, these bones began to grow muscle, organs, and flesh. Then an amazing thing happened: they came to life and became a massive army capable of winning large-scale conflicts. When Ezekiel asked God the meaning of this vision, he was told in vs. 11-14 that this was the future rebirth of Israel in the Promised Land, a prophecy that is being fulfilled as we live in our generation! For nearly 2000 years, since the destruction of Jerusalem by the Romans, Israel has been a "valley of dry bones". Then, during World War II, Satan once again tried to annihilate the Jews during the tragedy now well remembered as the Holocaust. God, however, in His infinite wisdom, "turned the tables" on Satan once again and from that event came a re-uniting of the Jewish people and a massive immigration of Jews to Israel. The land of Palestine once again saw the Star of David on its flag as Jews now began to move to their homeland in mind boggling numbers and there, in 1948, Israel was once again declared a nation. This event is pivotal to every other prophecy in this study and will soon be proven to be one of the central themes in this book, because it is one of the main signposts pointing to the fact that we are the last generation in the Church Age.

Following the prophecy of the valley of dry bones in Ezekiel 37, we have the great prophecy of the final battle of the age, perhaps the one that will set the stage for the Antichrist. This could occur just before the rapture, during the Tribulation Period, or just before the Second Coming at the end of the Tribulation. The important thing I want to consider here is the fact that even as I write this book and assemble this material, the nations of the world are coming together like pieces in a large puzzle, to align themselves exactly as shown in Ezekiel 38. So, in light of that statement, let's examine the chapter a little further.

Ezekiel is told to prophesy here against the following countries: Gog, in the Land of Magog; including their major cities of Meshech and Tubal; Persia, Ethiopia and Libya; Gomer, and finally, Togarmah. Where in the world are these places? And is this prophecy relevant to the nations that occupy these geographical spaces? It sure is! Gog is generally believed by Biblical scholars to be a reference to Russia while the land of Magog speaks of the larger area known as the USSR before the breakdown of Communism. The chief cities of Meshech and Tubal were the areas presently occupied by Moscow and Tobalsk in Russia today. Persia is the area known today as Iran and Iraq, Ethiopia and Libya retain the same names. Gomer is Germany, which brings out an important fact... Note that the prophecy does not say anything about East Gomer and West Gomer! It speaks of a united, reunited Germany, an event that has not only taken place in our lifetime, but just in the last few years!!! Down comes the Berlin wall, and with it another piece of the puzzle falls right into place in the prophetic scene. Finally Togarmah refers to Turkey (in the North Quarters). Even as I write these pages, Turkey is a hot-bed of controversy as the leaders of the nations fight for position in the Middle-East and deal with the instability in Iraq, with insurgents and terrorists trying to thwart the efforts of those who would try to secure freedom and peace in that area.

But why would these nations gather together to attack Israel? Certainly there is enough hatred by the Arabs in Iran and Iraq to foment this kind of mischief. But what about Russia, Ethiopia, Turkey and Germany; Why them? I think the answer is in one word: Economics. For years the Jews have been not only the most hated people of the world, but among the richest. No matter where they go, they tend to accumulate wealth and have done so for years. And with their return to their homeland, they have begun to take it back with them. Why do you think Hitler hated them so? Because of the threat they represented to his economy. They had corporately amassed so much wealth that Hitler and his leaders were afraid of losing their power to Jewish control. And the story is the same everywhere they go. The blessing of God follows them, even in exile!

With their return to Palestine, they are bringing back with them the technological secrets of the world, the wealth of many years of hard work, and the blessings of God. With the fall of Communism and the pending financial and social collapse of most of the Eastern Block countries, all eyes are beginning to look to Israel. And even more important is the fact that God told Ezekiel He would put this desire (to attack Israel) in the hearts of the rulers of these nations. Do you believe this? The Bible says that the nations in the north would come down upon Israel "To take a spoil, and to take a prey; to turn thine hand upon the desolate places that are now inhabited, and upon the people that are gathered out of the nations, which have gotten cattle and goods, that dwell in the midst of the land."[1] I didn't say this my friend. God said it. And neither you nor I have any reason to doubt it will happen and is already beginning as you read this book. Note by the way, the English translation of the word. Spoil ends in oil: spOIL! Oil can certainly play a big part in this. The world's economies are enslaved to that gooey black liquid. If you don't believe so, try going one day without using any. Every time you drive your car, you are using it. Every time you flip on a light switch you are using oil at the power station where your electricity is being generated. Yes, my friend, the nation that controls the oil controls the world. Oil represents a perfect opportunity for the Anti-Christ to bring the world to its knees during the days in which he will be setting up his one-world government. Also note that in the last decade, the 1990's, NATO and the United Nations have fought and won their first war (the Persian Gulf War) as a united world military power. Just a few short years ago! And what was that all about? What drew the United States and the Western European Alliance into the conflict? Oil, of course. Saddam Hussein had taken over Kuwait, a tiny nation that produces large amounts of the oil used by the west. Additionally, he was threatening to move into Saudi Arabia, perhaps the largest exporter of oil to the modernized nations of the western world. If ever there was a need for the member nations of the North Atlantic Treaty Organization (NATO) to act, this was it. But not without the United Nations! Every step of the process from the sanctions, to the actual war, to the inspections following the war and additional sanctions afterwards, was monitored by the United Nations. And now it's Bosnia, Serbia, and the former Yugoslavia, and Iraq that the UN is sticking its nose in. Are we getting closer to a one-world government? You bet we are!

What about the United States? Good Question. Would God simply abandon the nation that has done more to spread the Gospel around the world in this century than any other? Whose side are we on? Where do we fit in Bible prophecy? Well, to answer these questions we must start with the fact that because North America wasn't even discovered when prophecy was written, and because prophecies and sign gifts were directed at the Jew, not the Church[2], it stands to reason that God would speak to the Jew

---

[1]. **Ezekiel 38:12** - I will go to those once-desolate cities that are again filled with people who have returned from exile in many nations. I will capture vast amounts of plunder and take many slaves, for the people are rich with cattle now, and they think the whole world revolves around them!'

[2]. **Luke 16:29** "But Abraham said, `Moses and the prophets have warned them. Your brothers can read their writings anytime they want to.' 30"The rich man replied, `No, Father Abraham! But if someone is sent to them from the dead, then they will turn from their sins.' 31"But Abraham said, `If they won't listen to Moses and the prophets, they won't listen even if someone rises from the dead.'"

in geographical terms they could relate to. We do find some mention of the countries that would come, and although not specifically named, a vague reference to the U.S. might be seen where it speaks of Sheba, Dedan, Tarkish (countries of Western Europe), and "the young lions thereof". [3] (Possibly the U.S. and other North American countries.) While the United States has traditionally been an ally of Israel, we have no guarantee that will continue. Here's one possible scenario: The Antichrist could unite and control all of Europe, Asia, and the Arabic nations. The economic pressure could become so great because of America's dependence on foreign oil, that there would be a very good possibility that the U.S. would abandon Israel. Sorry, America; I wish I could tell you that we will be the good guys with white hats in the Tribulation, but that doesn't seem very likely from all indications.

In light of that statement, let's take a brief look at Daniel 2:36-45. Here Daniel is explaining to Nebucadnezzar the interpretation of the dream in which he saw a giant image like a man with a head of gold, body of brass, and legs of iron and feet of iron mixed with clay. This prophecy was to tell of how the world's nations and governmental systems would be set up in the future. Remember, at the time this was written, none of these things had happened yet. The head of gold referred to the Babylonian Empire. The body of brass was a reference to the Medio-Persian Empire yet to come. The legs of iron was a reference to the future Roman empire and the feet of iron mixed with clay were the yet to come Western Europe and their various allies (possibly including the U.S.) which would form the final governmental system. I personally believe that the ten toes speak of the major participants in the European Common Market alliance, something that has come to pass only in the last 30 years.

In all your studies of Bible Prophecy, while you may consider these last few items of interest, it is extremely important to note the following:

- Bible prophecy is always written for the Jew. While it may also have the secondary effect of enlightenment of Christians and warning of the lost, never forget that the Jew is the one to whom it is primarily written. Bible prophecy is God's love letter to His chosen people, Israel.

- That being the case, all major geographical references in Bible prophecy will generally be made with reference to Israel and the Middle East. It's not that God intends to leave out the Western nations in His dealings with man, but rather that He has chosen

---

[3.] ***Ezekiel 38:13*** - But Sheba and Dedan and the merchants of Tarshish will ask, "Who are you to rob them of silver and gold? Who are you to drive away their cattle and seize their goods and make them poor?'

to deal with the Gentile nations of the world through the Church. In this, the final days of the Age of Grace, the redemptive work of Jesus Christ is the only important thing that the Gentile nations of the world need to be concerned about.

- Most, if not all, major prophecies related to world governments and the alignment of the nations in the last days have already been fulfilled or are being presently fulfilled right under our noses in the generation in which we live right now.

In summary, let me say that God has a specific plan for dealing with man. It is two-fold, in that He deals with the Jew in one way, while He deals with the Gentiles in another, and yet opens the redemptive work of His Son to both in the Church. This plan was laid out in detail in the Old Testament Scriptures, and, looking back, we have the unique privilege of seeing every detail of it fulfilled, exactly as God said it would be. As the Psalmist said, "Selah!!!" (Think about this!)

CHAPTER 5:
"SEVENTY WEEKS"

In much of my previous discussion of Bible prophecy and its relation to current events I have made reference to the "Great Tribulation" or the "Tribulation Period". I have spoken of this as an event yet to come and mentioned it as casually as I would yesterday's news. I cannot and should not do so without offering at least some explanation of what this refers to. There is a prophecy in the Bible that goes something like this: "Seventy Weeks are determined upon thy people and upon thy Holy City to finish the transgression, to make an end of sins, and to make reconciliation for iniquity, and to seal up the vision and prophecy, and to anoint the Most Holy."[1] Even at a casual glance this seems to be a complex statement. Has it happened? Will it happen? If so, when? Well, to answer that I think we should look at the prophecy one piece at a time. First of all, we see that it speaks of a period of seventy "weeks" [2] However, when you look at the original Hebrew word translated "weeks" in the King James Version it actually means "sevens" or periods of seven. A more correct rendition of

---

[1] *Daniel 9:24-27* - "A period of seventy sets of seven has been decreed for your people and your holy city to put down rebellion, to bring an end to sin, to atone for guilt, to bring in everlasting righteousness, to confirm the prophetic vision, and to anoint the Most Holy Place. 25Now listen and understand! Seven sets of seven plus sixty-two sets of seven will pass from the time the command is given to rebuild Jerusalem until the Anointed One comes. Jerusalem will be rebuilt with streets and strong defenses, despite the perilous times. 26"After this period of sixty-two sets of seven, the Anointed One will be killed, appearing to have accomplished nothing, and a ruler will arise whose armies will destroy the city and the Temple. The end will come with a flood, and war and its miseries are decreed from that time to the very end. 27He will make a treaty with the people for a period of one set of seven, but after half this time, he will put an end to the sacrifices and offerings. Then as a climax to all his terrible deeds, he will set up a sacrilegious object that causes desecration, until the end that has been decreed is poured out on this defiler."

[2] *The Schofield Reference Bible*, King James Version (1611) Oxford University Press, 1/1/1909, p 914

the verse, such as that used in the New Living Translation, which we have used for our footnotes, would therefore be "Seventy Sevens (of years) are determined...." What about the result? Has the transgression been finished? No. Has sin ended? Again no. Of course it hasn't. It's all around in us and even still in us. What about making reconciliation for iniquity? Yes. That was done nearly 2000 years ago at Calvary. How about "seal up the prophecy". Again yes. That happened with the completion of the written text of the Bible, ending with the last chapter of Revelation. Finally, how about "to anoint the Most Holy"? No. That is yet to come when the final "week" is completed and Christ sits on the throne of David during the Millennial Kingdom. So there you have it: two yes and three no. This means that we must still be somewhere between the beginning and the end of the "Seventy Weeks" prophecy.

In his book "Rightly Dividing the Word of Truth", Dr. C. I. Schofield does an excellent job of detailing the course of events that occurred during the first 69 "weeks" of the prophecy. Here is a brief synopsis of what he discovered which will help us to better understand the meaning of the prophecy:

In the Bible, when reference is made to a year, it is not what we know as a solar year of 365.25 days. Look at Genesis chapter 7 and chapter 8, the story of the Great Flood and Noah and the Ark. The worldwide flood started in second month on the seventeenth day and ended on the seventh month and the seventeenth day when the ark came to rest on dry land in the mountains of Arrant in what is now Turkey.[3] That's five months, exactly. We're told that "after the end of the one hundred and fifty days the waters abated"[4].

You need not be a rocket scientist to know that 150 divided by 5 is 30. So, when dealing with Bible Prophecy, you should always assume that we are talking about 30 day months or 360-day years. In light of this let's look a little closer at the Seventy Week prophecy. It starts with the going forth of the commandment to restore and rebuild Jerusalem. This was the commandment issued by King Artaxerxes to rebuild the wall as recorded by both Ezra and Nehemiah. It continues to the triumphal entry of Christ into Jerusalem on what we know as Palm Sunday. Then, as the prophecy states, Messiah is "cut off". This is a reference to the Crucifixion, Death, Burial, and Resurrection of Christ. This period of time covers 434 years, each 360 days long, (exactly to the day!) or sixty nine of the seventy weeks.[5]

[3.] ***Genesis 7:11*** - When Noah was 600 years old, on the seventeenth day of the second month, the underground waters burst forth on the earth, and the rain fell in mighty torrents from the sky.

[4.] ***Genesis 8:3-4*** - So the flood gradually began to recede. After 150 days, 4exactly five months from the time the flood began the boat came to rest on the mountains of Ararat.

[5.] ***Rightly Dividing the Word of Truth*** (C. I, Schofield) Alva J. McLain, 1965

Then comes the destruction of the city (Jerusalem), a well documented historical fact.[6] The destruction was not only foretold here, but we are even told who would do it nearly 500 years before it happened.

The text mentions the "people of the Prince that shall come" a reference to the Anti-Christ who would be revealed during the seventieth week. He will come, as mentioned in the previous chapter of this book, from a new "Roman Empire" or as I stated earlier, an alliance of the western nations of the world, most of which hadn't even been discovered yet when this prophecy was written. Following the destruction of Jerusalem comes a long pause, not unusual in Bible Prophecy. This is what I have made previous reference to as the Day of Grace or the Church Age. Finally, when the time is right, comes the Seventieth Week, the final seven-year period known as the Great Tribulation. We find reference to this also in Jesus' teaching in the Olivet discourse.[7] and the majority of the book of Revelation is also filled with details about what will happen during this period of time.[8]

Has this Great Tribulation occurred yet? If so, when? If not, when will it happen? Could it possibly happen in my lifetime? And will we, (the church) be around when it does? The first answer is no. It has not happened yet. There are numerous plagues recorded in Revelation, which will happen during that time. We have no record of these having ever occurred. Also,

---

[6] *Daniel 9:25-27* - Now listen and understand! Seven sets of seven plus sixty-two sets of seven will pass from the time the command is given to rebuild Jerusalem until the Anointed One comes. Jerusalem will be rebuilt with streets and strong defenses, despite the perilous times. 26"After this period of sixty-two sets of seven the Anointed One will be killed, appearing to have accomplished nothing, and a ruler will arise whose armies will destroy the city and the Temple. The end will come with a flood, and war and its miseries are decreed from that time to the very end. 27He will make a treaty with the people for a period of one set of seven but after half this time, he will put an end to the sacrifices and offerings. Then as a climax to all his terrible deeds, he will set up a sacrilegious object that causes desecration, until the end that has been decreed is poured out on this defiler."

[7] *Matthew 24:15-26* - "The time will come when you will see what Daniel the prophet spoke about: the sacrilegious object that causes desecration standing in the Holy Place"--reader, pay attention! 16"Then those in Judea must flee to the hills. 17A person outside the house must not go inside to pack. 18A person in the field must not return even to get a coat. 19How terrible it will be for pregnant women and for mothers nursing their babies in those days. 20And pray that your flight will not be in winter or on the Sabbath. 21For that will be a time of greater horror than anything the world has ever seen or will ever see again. 22In fact, unless that time of calamity is shortened, the entire human race will be destroyed. But it will be shortened for the sake of God's chosen ones. 23"Then if anyone tells you, `Look, here is the Messiah,' or `There he is,' don't pay any attention. 24For false messiahs and false prophets will rise up and perform great miraculous signs and wonders so as to deceive, if possible, even God's chosen ones. 25See, I have warned you. 26"So if someone tells you, `Look, the Messiah is out in the desert,' don't bother to go and look. Or, `Look, he is hiding here,' don't believe it!

[8] *Ibid* – See Chapter 2, Note 7

the Anti-Christ will have to stand in the Temple at Jerusalem and declare himself as God; a temple that hasn't even been built yet. By the way, it has been revealed in the last couple of years that engineering studies show that the original Temple site was not where the "Dome of the Rock" now stands as had been believed for many years, but a few hundred feet away where there is now a small building. Don't let it surprise you if you hear that plans are presently being drawn to start the reconstruction of the Temple.

There is one interesting twist to this however. All this must come to pass in "this generation" according to Jesus.[9] Many believe this to be a reference to Israel's preservation unto the times when all these things are fulfilled. This is certainly true. But any other time the word generation is used in the Bible, it usually refers to a specific set of years during which a specific person lived. Example: "the generations of Adam". [10] What is a generation then? How long is it? When did it, or when will it start? We'll get to a detailed discussion of that subject in a later chapter. But first, let's look at some other recent developments as we put more of this puzzle together, piece by piece.

Before concluding this chapter, let's talk about the Church. Where will it be during the Tribulation? Since this is a question that has not been answered directly word for word, it has long been a cause for dissension among those who study Bible prophecy. There are some who believe the Church will go through the Tribulation. There are others who believe, as does this writer, that we will be removed in an event called the "Rapture" before it starts. And there are even some who believe that the Church will be removed "mid-trib" or during the middle of it. Those who believe as I do are called "pre-millennial, pre-trib" believers because we not only believe that the Church will be removed from the earth before the Millennial Kingdom, but also that it will be "taken up" before the Tribulation Period. There are several places in which we see this, when Scripture is taken in context. (Remember our rule of interpretation in the beginning of chapter 2 in this book?) Always keep Scripture in context. II Peter 1:20-21 tells us to do so where it states that "no prophecy of the Scripture is of any private (separate) interpretation". Never take a verse by itself and build your theology on it. (Unless you want to start a new cult and we have enough of that around already!)

[9.] ***Matthew 24:33-34*** - Just so, when you see the events I've described beginning to happen, you can know his return is very near, right at the door. 34I assure you, this generation will not pass from the scene before all these things take place.

[10.] ***Genesis 5:1-2*** - This is the history of the descendants of Adam. When God created people He made them in the likeness of God. 2He created them male and female, and he blessed them and called them "human."

So let's look at a few passages of Scripture in their surrounding context and see what they have to say. Among the first things I would like to consider is the fact that there was some apparent concern on Paul's part that the Thessalonians may have thought that the Day of Christ (the Second Coming) had already happened and that they may have missed something.[10] Paul writes to tell them that some things must happen first[11] These include: a.) A "falling away first" and b.) The Man of Sin (Antichrist) is revealed. While many believe the falling away here to be a worldwide apostasy just before the Tribulation, there is also a popular view that this makes reference to the rapture. The Greek word used here can also denote a "catching away" or "removal from" which would clearly line up with the Rapture. First comes the "catching away", then the revelation of the Man of Sin.

Another excellent text for the Rapture is in I Thessalonians 4:13-18[12]. Here, we are told of an event in which the dead shall rise from their graves, unopened, be reunited with their soul and spirit, then meet with those who have not yet died and with the Lord in the air. Note that in this passage, Christ does not actually set foot on the earth, therefore this cannot be the Second Coming described at the end of the book of Revelation. And while we are talking about Revelation, remember our discussion in chapter 2 about the seven churches in Revelation 1-3. After the last church leaves the scene, John is caught up into Heaven, and the Tribulation begins. This is a perfect picture of the Rapture of the Church.

Finally before closing this discussion I must not fail to point out that we are told "God hath not appointed us unto wrath"[13], a clear indication that the Church will escape the Wrath of God to be poured out with a vengeance upon an unbelieving world during the Great Tribulation. I believe that we are closer to the beginning of the Seventieth Week of the prophecy in Daniel than any generation that has preceded us and that our generation may, in fact, be the last generation to enjoy the peace and safety that God's patience has brought with the Church Age.

[11.] *2 Thessalonians 2:1-4* - And now, brothers and sisters let us tell you about the coming again of our Lord Jesus Christ and how we will be gathered together to meet him. **2**Please don't be so easily shaken and troubled by those who say that the day of the Lord has already begun. Even if they claim to have had a vision, a revelation, or a letter supposedly from us, don't believe them. **3**Don't be fooled by what they say. For that day will not come until there is a great rebellion against God and the man of lawlessness is revealed--the one who brings destruction. **4**He will exalt himself and defy every god there is and tear down every object of adoration and worship. He will position himself in the temple of God, claiming that he himself is God.

[12.] *Ibid* – See Chapter 2, Note 6

[13.] *I Thessalonians 5:9-10* - For God decided to save us through our Lord Jesus Christ, not to pour out his anger on us. **10**He died for us so that we can live with him forever, whether we are dead or alive at the time of his return.

CHAPTER 6:
SETTING THE STAGE

"Many shall run to and fro and knowledge shall be increased"- **Daniel 12:4**

One of the major attractions at Walt Disney World in Orlando is a short ride through a beautiful little group of small work colonies featuring little people of all nationalities and languages working at their daily tasks. The attraction is called "It's a Small World" As you pass by one group after another you hear the theme song "It's a small world after all" sung again and again, sometimes in different languages. It is as though you could visit the entire world in fifteen minutes! We live today in a small world. I wonder what it would have been like in the 18th or 19th century to attend a business meeting of a large shipping company? Perhaps there would be some discussion about which ship would carry the cargo where and how many weeks it would take to cross the Atlantic or Pacific. What a comparison to today. That same meeting would probably be a satellite teleconference in which Europe, China, Japan, and American businessmen are all seeing and speaking to each other without anybody leaving their office. They would be talking about how many hours it would take to move merchandise from one part of the world to the other instead of weeks.

With the discovery of electricity in the 18th century, the telephone in the 19th, air travel and electronics in the 20th, we have shrunk the size of the world to the point where we now can consider someone halfway around the globe "neighbors". The start was with telephone communications. First came the crude telephone devices put together by Alexander Graham Bell that allowed people to talk across the street or across the town. Then, with Marconi's discovery of radio frequency transmission came the wireless telegraph, the grandfather of radio communications, as we know it today. In the early 1930's radio stations started to pop up all around the U.S.

using amplitude modulation (AM) many of which are still in existence today. Telephone communications began to become a worldwide business with the laying of the Trans-Atlantic telephone cables by AT & T. The 1950's saw improvements in radio, making signals clearer and static free by use of Frequency Modulation (FM), which produced a quality of sound and music transmission second to none. We now are developing direct transmission of Digital Audio Broadcasting (DAB). It is already available on L-Band for radios equipped with satellite receivers, and will soon be available as a companion to your favorite AM/FM radio station. The technology for "IBOC" digital (In-Band-On-Channel) transmission of digital audio is currently being installed in many radio stations. As Chief Engineer for WNSB-FM, I have had part in designing the equipment package that will bring IBOC to our station, sometime around the fall of 2005.

Television revolutionized the way entertainment services are delivered. Could our grandparents, back before World War II, have ever imagined seeing someone speak to you on a screen in your home? And in full color with stereo sound! Do you remember the time when you had to go to a movie theater to do that? Some of you do. Now we have a whole new industry called the "Home Theater Business" providing big screen pictures with Surround Sound in your living room!

Microsoft has introduced interactive television allowing you to do your banking, shopping, schooling, go to work at home and perform countless other tasks without ever leaving the comforts of home, all using your TV set and a telephone line. Data processing has so infiltrated the home that most major companies now offer residential ISDN (high speed data) lines to your home. And Phillips Corp. is now selling an Internet converter that makes your regular television an Internet computer, called "Web TV". What ever happened to the days of sitting around the family table with a Monopoly game or a 5000-piece picture puzzle?

With the discovery of Space Travel in the 1950's and 1960's came the first placement of satellite devices in geosynchronus orbits around the earth. Now we have a "Bird Line" of satellites above the equator each carrying thousands of channels of radio, television, navigational and communications channels. Some are so powerful you can point a 0.5-meter dish (approx. 1') at them and pick up hundreds of television channels. The transatlantic telephone cables have become obsolete. Most long distance telephone communications are relayed from point to point by satellite. We have a global positioning system (GPS) now in use. Ships use it instead of the good old-fashioned compass.

And what about your car? We no longer have mechanics at the dealerships. We now have technicians and master technicians, and rightly so. Automobiles today are rolling computers. The computer supervises everything the car does from emissions to cruise control. Onstar _{tm} keeps your vehicle in constant communication, by satellite, with a live operator you can contact with the push of a button. You can get help in an emergency, assistance with directions, or service during a mechanical failure. And now we have keyless entry, remote start, and even totally keyless cars! Just put the radio-controlled fob in your pocket; Walk up to the driver's door and it unlocks automatically; walk away from it and it locks. Sit in the car, and just push the start button: no key needed! And if you lock the "smart key" in your trunk, it automatically pops open to warn you that your key is locked inside.[1]

Many cars use a system called the "LoJack" theft locator to find where your car is when it gets stolen. We are already conducting tests with "smart highways" where you get in your car and the computer locks into the GPS network and data transmission equipment in the roadway and drives you where you are going while you read the morning paper. Radio stations are operated unattended, from points far from them by remote control using programs and control signals delivered to them by satellite.

In the late 1950's, as a budding teenager, I used to look forward to my almost daily trips to the local Boy's Club of America in my home city where I would learn how to swim, play games, and most important, use a typewriter. The typewriter.... What a fantastic invention! The Palmer Method of handwriting (which I hated) was now history for me. I was a big shot.. I could type. Today, at age 58, I have a 24-year-old daughter who has been typing since she was 5. She was using a word processor at age 6 and working databases by the time she was 7. I have often heard it said, "If you want to know something about computers, ask your kid." In the movie Star Trek VI, two middle aged leaders, Captain Kirk and Chancellor Gorkon, a Klingon, are having a discussion about an upcoming alliance between the two races in which Gorkon tells Kirk, "If there is to be a brave new world, our generation will have the most difficulty living in it."[2] My generation certainly can say that. Many folks in their 50's and 60's today are resistant to the intrusion of computers into their lives and wouldn't have one if it were given to them free! Those who are willing to accept them often do so reluctantly. The younger set on the other hand, the "baby boomers", seem to embrace the newer technologies with open arms. I wonder what my great-grandfather would do if he had to come back to life and live in today's world? Panic, probably!

Computers are only the beginning. When we entered the 20th century, horse drawn carriages were still the most popular means of transportation.

[1]. **MSN News** www.msn.com 4/20/2005

[2]. **Star Trek VI** (the Movie) Paramount Studios 12/06/1991

Since then, in only the last 100 years, we have gone from horses to cars that now travel faster than a horse can imagine, to air travel that can take us around the world in one day to space travel that can take a man to the moon in one week. The Scripture says that in the last days that "man shall run to and fro about the face of the earth and knowledge shall be increased"[3] Take a look at the world of medicine and health. In 1900 people routinely died from colds and influenza. Remember Polio in the late 1940's and early 50's? Here's a disease that is almost unheard of today. We have learned more about the makeup of the human body, in the last 100 years than in the previous 1900 combined. We now know how the brain communicates with every other body part via a network of neurons and synapses through a susbstance called Seratonin. How well the Seratonin works controls everything from emotions to mental health. From that knowledge come all types of drugs that control depression, anxiety, weight control, and many other bodily functions.[4]

We talked earlier about this "Great Tribulation" period, which most Bible Scholars predict will come in the near future. I would like to examine here the relationship of current technologies to some of the specific occurrences that will take place during that time. Some of these will, in fact, verify that it is a future event. It could not have occurred in the past and however close we may be, it is not occurring now. One area that bears careful examination is world trade and the financial system.

Not long ago, I stood in a checkout behind a frustrated customer who had to walk away and leave his merchandise because his check was rejected by the scanner located at the cash register. What had happened was that he had overdrawn his account on a recent occasion and his name and number went into a database set up to identify potential bad check writers. In ten seconds a computer, miles away from the store, without the assistance of a human operator, made a decision not to sell this person any merchandise. When I was born, such a system of buy-sell control hadn't even been imagined. The prophecy in Revelation 13 was a mystery. Today the system by which that will occur is in place at virtually every checkout counter in America and throughout most of the industrialized world. You can make transactions with your bank at millions of ATM locations. (If the "beast" (ATM machine) lets you, that is!)

Friend, we are not just close to worldwide buy-sell control by one government or one person. we are right at the doorsteps. The technology is *already* in place. Then there's the Internet. Talk to anybody, anywhere, in the entire world without ever leaving your home. Shop anywhere, use

---

[3] ***Daniel 12:4*** But you, Daniel, keep this prophecy a secret; seal up the book until the time of the end. Many will rush here and there, and knowledge will increase."

[4] ***Time Magazine*** Oct 1997 Issue

any library, attend any school, the possibilities are limitless. And with them comes the menacing possibility that one person can know everything there is about you: Your age, health, wealth, buying habits, everything is public knowledge today. Your name is on hundreds of mailing lists. Been to your mailbox lately or checked your e-mail? Those Internet "cookies" are little databases that collect information about you and store it in a "temporary" file on your computer. Anyone who wants to know what sites you visit, where you shop, what you buy, or any personal information you have shared with other sites can read them with the proper software while you are on-line.[6] In this electronic age, often called the Information Superhighway, we have traded our personal privacy for the convenience of the Internet. (I personally recommend deleting all "cookies" frequently!)

And what about the world financial system? Have you noticed how worldwide currency values are all referenced to the dollar as the standard? In order for the antichrist to have control of the world he will have to control the currency. There will need to be a one-world money system, electronically controlled, electronically transferred, with no paper. Paper is too slow. When you write a check, it is usually deposited within twenty-four hours and funds are deducted from your account and electronically transferred that same evening even though it may take days to get the actual check back to your bank. That is why it takes several days to get your bank statement at the end of the month or if someone writes you a bad check it takes several days to get it back to you. The checks have to catch up with the transactions. Eventually the paper will be eliminated. In fact, practically all of the larger grocery chains have electronic payment terminals at the cash register. Just swipe the card and payment is made without cash or check, electronically, and your checking account is debited immediately.

A recent development is the surgically implanted microchip often used in pet identification. It is so small you can barely see it, yet it can contain everything there is to know about you, including your account information, etc. and can be read and updated with a pocket size scanner. I recently read an article in C-Net news that appeared in the November 26th 2003 edition of MSN's home page. It tells of a company that has now improved this technology, making it possible to inject one of these identification microchips under human skin by using a small syringe without even making an incision! The technology still needs a few security issues resolved, but it's here.[5]. The Bible tells us that there would be a time when there will be one world leader who will have in place, a system that can control who buys or sells anything. "And he causes all, both small and great, rich and poor, free and slave, to receive a mark on their right hand or on their

---

[5] **C-Net News** www.msn.com 11/26/2003

foreheads, and that no one may buy or sell except one who has the mark or the name of the beast or the number of his name."[6] Ten years ago this was a prophecy that was still unsolved. Why would anyone want to have a number on his hand or forehead? Well, now we know. During the Great Tribulation, the means of identification will most likely be this implanted microchip and those left after the rapture will find it a necessity to have one and pledge a commitment of allegiance to the world leader to get it.

In addition to one-world currency standard and electronic money, there will need to be a one-world system of measurement. Have you noticed the subtle entrance of the metric system into your life? The bolts and nuts on your car are no longer fractional sizes, even if your car is American made. They're metric. A mechanic, who does not own two sets of tools today, metric and American, is an extinct species. Automotive technicians now have to be familiar with the onboard computer that controls everything from the engine to the fuel tank. Just wait until your car shuts down for no known reason while you're driving down the road one day and you'll see why. Mine recently did and it took $10,000 worth of diagnostic equipment to find that some battery acid had leaked down onto a cable harness and the resulting corrosion was creating a false signal to the engine shut down relay. In many parts of the country, gasoline is metered in liters, not gallons. Speed limit signs have both miles per hour and kilometers per hour. You've seen them. You're no stranger to what I'm talking about.

To say the stage is being set is an understatement. It is not only being set, but it is being set exactly as the Bible predicted right "under your nose"! And (as the theme of this book goes...) *in our lifetime!* Are you ready for the "Mark of the Beast"? If you don't leave this earth before the seven-year Great Tribulation, you will probably get one. Open your heart to Christ today and you won't have to be around for any of this.

---

[6] **Revelation 13:17-18** - And no one could buy or sell anything without that mark, which was either the name of the beast or the number representing his name. **18**Wisdom is needed to understand this. Let the one who has understanding solve the number of the beast, for it is the number of a man. His number is 666.

CHAPTER 7:
<u>666</u>: THE COMPLEXITY OF A SIMPLE NUMBER

A third grade teacher was giving her students a small spelling quiz. She told them, "For your assignment today, I want you to tell me the name of the work either one of your parents does and spell the word and name the place where they work and spell that word. Who will be first?" A little girl on the first row stood up and said' "My dad is a lawyer – That's L-A-W-Y-E-R and he works in court – that's C-O-U-R-T." "Very good." The teacher replied. "Who will be next?" Little Johnnie stood up and said, "My mom is a nurse – that's N-U-R-S-E and she works in a hospital – that's H-O-S-P-I-T-A-L. "Excellent" said the teacher. Billy was next so he stood up and said, "My daddy is a chiropractor – that's C-H-A-P-R – uh I mean – C-H-I-P-R..." The teacher interrupted and said, "Well, Billy, that's a big word for a third grader; have a seat and we will get back to you in a few minutes." Next up was Tommy. He jumped to his feet and said, "My dad is a bookie and I'll bet you $20 Billy can't spell chiropractor!"

Numbers, numbers, numbers. The world runs on them. From accountants and bean counters to bookies and loan sharks, everybody uses them. Mathematicians spend most of their lives learning how to teach them to other people, scientists try to unscramble them to unlock the secrets of the universe. Kindergarten children play with them to find out what they are. If you are educated well enough to be able to read this book, then you have probably learned how to use them as well, for everything from determining the speed you are driving to calculating the balance in your bank account after you have made your weekly contribution to the IRS (or whatever the taxing authority is in your locale).

Just out of sheer curiosity, I made it my personal project to ask most of the people I came in contact with one day recently, if they had ever heard of the number 666 and what it meant. Surprisingly, most of them made some kind of association between the Antichrist and the Beast of Revelation chapter 13 in the Bible and some were even aware that it would be some kind of number that would be required to buy or sell anything when the Antichrist comes into power. And they were right. Here's what the Bible says about that:

> "And he causeth all, both small and great, rich and poor, free and bond, to receive a mark in their right hand, or in their foreheads: And that no man might buy or sell, save he that had the mark, or the name of the beast, or the number of his name. Here is wisdom. Let him that hath understanding count the number of the beast: for it is the number of a man; and the number is Six Hundred Three Score and Six." *Revelation 13:16-18 (King James Version)*

So, does this mean that everybody will go around with 666 tattooed on their hand or forehead? I doubt it. Most people would see right through that as my little survey indicated. It would have to be something subtler that would creep in and seem to be a necessity to keep pace with society. In chapter 5, I mentioned a new technology that would make it possible to have a personal recognition data chip injected painlessly under the skin, similar to the electronic pet identification systems in use now. It would allow a user to access his bank accounts at the ATM's or point of sale systems we now have in place. Not only is this technology a recent invention (within the last two years – and it is still undergoing some refinement and fine tuning) but it also shows how the prophecy we just quoted from Revelation 13 can easily be fulfilled *in our lifetime!* Prior to the introduction of this technology, along with the whole point of sale system, with its bar code scanners and universal product labels, there was no way to make any sense of this prediction.

Now the question looms: What does 666 mean, how does that particular number relate to our current financial system, and why the choice of that combination of numbers? Well, I can only speculate on the details of how it would work, but I can be firmer on the fact that it can work, *in some form*, with today's high-speed computer networks including the Internet, the worldwide ATM system, and the worldwide point of sale system. So let's toss out something here for your consideration:

In order to understand how this might work, it would be helpful to give you a short primer on the basics of how a computer works. Just prior to the introduction of the earliest computers, communications was primarily transmitted and received over either voice grade telephone lines, by radio systems of one type or another, and dating back to the earliest computer-like systems: teletype machines. Nearly unheard of nowadays, these monsters were still clicking away at a whopping 60 words a minute when I began to study them and work with them in the military during the late 60's. We even had one that had some type of on-board memory system that would receive and store the data much faster, print out several lines at once, pause for a second to receive another burst of signals, then print another 10 lines or so. When it printed, it sounded like a loud burp or belch, so we nicknamed it the "Burpee".

So what was a Teletype signal and how did it work? Remember, a computer knows only two numbers: one and zero, ON or OFF. In fact, most modern computers use a 5-volt data bus supplied by a voltage regulator in the power supply. Lose this and it's bye-bye computer. This is called binary data. And like binary code, teletype machines used a binary sequence consisting of a start bit, six data bits, and a stop bit. We would often use the letters R & Y to run continuous test patterns because the binary code for R was 101010 and for Y it was 010101, with the start and stop bits added to the sequence. We also had another favorite test phrase, "The quick brown fox jumped over the lazy dog's back 1234567890 times" which used every letter in the alphabet and every number at least once.

From these crude machines came modern computers, the first of which, like their predecessor the TTY machine, were 8 bit machines. But as computers became smarter and more definition of characters, and especially colors, was needed, they expanded to 16 bit machines. The Z-80 and 8088 processors were the eight bit standards and were followed by the 286 thru 486 boxes, which were sixteen bits. Then along came the Intel Pentium microprocessor chips. These 32 bit systems are what most machines use today. Look in your Windows directory and you will see a folder called System 32. This is where your operating system stores all the 32-bit information that is associated with various characters, symbols, musical notes, and colors. For this reason your computer can now have up to 256 colors, play music, and even employ voice recognition, as software gets smarter and computers get faster. We're talking stuff that was science fiction in the 1985 Star-Trek television series, and is now reality, today, only 20 years later!

So how does all this relate to 666? Well suppose you took three lines of six numbers:

865478
347890
345679

One problem: Using the base-10 numbering system that we are all so familiar with, there would be severe limitations as to how much data you could squeeze into three lines of six character digits. Although possible, you would have hardly enough room to assign a unique number to every one of the world's 6 billion men, women, and children, much less individual accounting and lifestyle and location data. But computers don't use a base-10 numbering system, where the count starts at one and ends with 10. They use a hexadecimal numbering system, where the count starts at one and ends at F:

1 2 3 4 5 6 7 8 9 A B C D E F

Lets just use a small security panel as an example. It notifies fire and police departments of emergencies using a 4+2 format, which takes only a couple of seconds to transmit and activate dispatch of emergency personnel and equipment. This consists of a four-digit account code and a two-digit event identifier:

1643 8F

The two-digit identifier would be limited to 99 events with base 10 codes. But with hexadecimal encoding, there are literally hundreds of possibilities. So now let's look at our sample "Number of the Beast" suggestion with hexadecimal numbers:

15FD6C
A78EC2
FF5E63

See what I mean? The possibilities of combinations are nearly limitless. With three lines of six numbers, the system could store endless amounts of data on each man woman and child. One line could match the user's name in a worldwide database, which would bring up records with personal information. The second line could track the person's location at any given time. The third line could be used to track his activities and identify his allegiances and religious affiliations. Or the three lines of hexadecimal code could somehow be used with the cookies or other hard coded information (your computer's IP address for example) to identify you, where you are, and what you're doing. Lookout... Big Brother is watching you! (George Orwell: 1984)

I don't know what method or format the Antichrist and his leaders will use. But what I'm trying to convey here is that we have the technology in place to make this "666" system of buy sell control a reality and it, like many of the other technologies we've already discussed has been developed only during the last 50 or so years, *in our lifetime!*

Remember.... "So likewise ye, when ye shall see all these things, know that it is near, even at the doors" The words of Jesus Christ, Matthew 24:33

CHAPTER 8:
THE TIMES OF THE GENTILES

Now there's a strange term. Who or what are Gentiles and what or when are their times? What is supposed to happen during the "Times of the Gentiles" and where does this term come from? It is a subject worth examining, since Christ mentioned it in the Gospel according to Luke. He said, "Jerusalem shall be trodden down of the Gentiles, until the Times of the Gentiles be fulfilled."

Before I go any further with this subject, I cannot assume that everyone who reads this book will know who or what "Gentiles" are. In the Jewish culture, there were only two groups of people: Jews and Gentiles. The Gentiles were generally considered to be non-believers, people of non-Jewish origin. Gentiles were usually polytheistic, meaning they worshipped multiple gods, while the Jews were monotheistic; meaning they only worshipped the one true God. Among the major Gentile cultures during the writing of the New Testament, were the Greeks and the Romans, both of which had gods named after all the major planets and stars. The Jews, as we mentioned earlier were God's chosen people, Israel. They were the ones through whom the Messiah was prophesied to come. However, they would also be the ones to reject Him as their Messiah when He did come. Why would they reject their Messiah, for whom they had so long prayed and waited? An understanding of the "Times of the Gentiles" will clear that issue and further confirm that He was indeed their Messiah. Except for a small minority of the Jews who would see the real reason for the Messiah's coming, the nation of Israel, as a nation, did not believe or trust Him because their visions of who Messiah is and what He was to do had degenerated from an expectation of deliverance from sin, to an expectation of physical deliverance from their rulers, who at the time of Christ happened to be Rome.

Since the phrase "times of the Gentiles" appears to denote a specific period of time, we should examine what the parameters that define that time are. Here's the main one: "Jerusalem shall be trodden down of the Gentiles" That would seem to indicate that it began when the Jewish people, the Nation of Israel, would lose its capitol, Jerusalem, to rule by another nation or people. It therefore must end when Jerusalem would once again be under control of Israel. There are several possibilities as to when this started. Some commentators teach that it started in 70 A.D. when Israel was conquered by Rome, and the Jews were scattered among the nations of the world. Some teach that it began with the captivity of Judah under Nebucadnezzer, (2 Chronicles 36: 1-21). More important to our study is the question, "When will it end?" Why? If it has already ended, or is about to, then we are in the last days for sure. So, with that thought in mind let's examine some specific possibilities. Remember, our main definition for the "Times of the Gentiles" is that Jerusalem will be under control by Gentile or non-believing, non-Jewish nations. Therefore, several things are important to consider here:

1. There will be no significant occupation of Jerusalem or most of Israel by the Jews during the times of the Gentiles. Arabic people who were descendants of Abraham's illegitimate son, Ishmael, would occupy the land of Palestine. The land would be under the political control of a Gentile nation (most recently Great Britain).

2. There will be no central Jewish government or Israeli military forces during this time. No political structure in the land of Palestine or anywhere for that matter would be present that is controlled by Jews.

3. The Gospel of Jesus Christ will have changed from being spread by a handful of Jews to being spread worldwide by Gentiles. The true Church will now be worldwide and God's gift of eternal life through Jesus Christ will be available to all people, Jews and Gentiles alike. (Please note that it always has been, but in the Old Testament economy, a Gentile was converted when he gave up his polytheism and turned to the God of the Jews and was saved by faith in the Messiah who was promised.)

4. The Jews will have barely survived a major attempt to annihilate them. Some believe that this reference, taken from Luke Chapter 21, in a statement by Jesus just before His mention of the "Times of the Gentiles" is a reference to the Great Tribulation. It also has happened during World War II under Adolph Hitler where over six million Jews were slaughtered like animals only because they were Jews and for no other reason. But I think that when you take Scripture in context, (Remember our primary rule of interpretation!),

you will conclude that this spoke of the conquest by the Roman army in 70 A.D. It is here that the armies of Rome surrounded the City of Jerusalem, exactly as predicted by Daniel, and "one stone was not left standing on another", exactly as predicted by Jesus. The Jews were driven out of Palestine and scattered among the nations, never to return until the "times of the Gentiles" would end. And we are the generation that is watching it draw to a close.

Now, that's exciting. It's exciting because it has happened *in my lifetime*. We are living in what God's wonderful Word calls the "last days". In 1948, Great Britain gave the land of Palestine back to the Jews, and for the first time in nearly 2000 years, the Star of David, on the flag of Israel, proudly flew over the land. But there's more, and it gets even better. In 1967, during the now famous seven-day war, Israel took over the West Bank, *including the City of Jerusalem*. Jerusalem was no longer under the control of the Gentiles. However, this is only temporary. Once again, during the yet-to-come Great Tribulation, the Jews will be run out of Jerusalem and will have to flee to the mountains. So, then, the "Times of the Gentiles" have not fully ended. It will end with the Desecration of the Temple, spoken of by Daniel the Prophet and referred to by Jesus in the Olivet discourse.

We can, however, conclude from this study that the "Times of the Gentiles" are, as already mentioned, nearing the very end. Israel is now a nation in her own land. Jerusalem is under her control. Jesus is coming soon. The rapture could be today. Are you ready?

CHAPTER 9:
100 YEARS

In 1800 mail was the primary means of communication between people who lived in different parts of the country. In the United States the U.S. Postal Service was the way people moved their letters. The Postal Service offered the "Pony Express" the fastest way to get mail to the West. We may laugh at that now, but just remember, that was less than 200 years ago! Then came the early 1900's. The telegraph, the telephone, and motorized transportation were all now in place. Instead of weeks to send a letter it now became a matter of days. Railroads moved things across the U.S. quickly. Then came airplanes. How many of you are old enough to remember the Airmail stamp issued by the Postal Service? Instead of the regular 3-cent postage it was 12 cents to get it across the country in 2 - 3 days. Today we have Fed-Ex, Airborne, DHL, UPS, RPS, Express Mail, (Did I forget anybody?) and a host of other SPD's (Small Package Delivery Services), all offering next day delivery, and some, at the right price, can deliver it the same day! FAX machines are a household item. When I was in the Navy only the military had them. The fax machine on my ship was a crude device about the size of a 21" TV set, with real wide thermal paper, printing out fuzzy, barely useable weather maps. And don't forget e-mail. With e-mail you can send and receive letters in a matter of minutes over the Internet. All in the last 100 years!

Medical science has even outpaced communications. Today we have medical imaging devices unheard of ten years ago. CT Scans, MRI (Magnetic Resonance Imaging), Laser Surgery, Nuclear Medicine, and Ultrasound are available at most hospitals and we are just beginning. Medical electronic technology is breaking new ground every day. Recently I was diagnosed with Prostate Cancer. The diagnosis was the result of a relatively new test called the PSA, (Prostate Specific Antigen) unheard of 30 years ago. An Ultra-Sound Biopsy confirmed the results. Surgical removal of the prostate gland eliminated the cancer from my body and today I am 100% cured and can expect a normal lifespan. Just 30 years ago, in the absence of these modern detection methods, men who had what I had usually died within 5 - 10 years after diagnosis. By then it had usually progressed past the point of "early detection & treatment". Using Ultrasound technology today, pregnant women can see images of their baby; even determine its sex, several months before birth. This helps couples make more intelligent preparations for the baby's arrival. Genetic science is discovering new information every day that will lead to the cure of many major diseases. Animal cloning has already started and soon it will not be too unlikely that it will be attempted with humans. The most recent tests being conducted are with pigs. Scientists are learning how to re-write a pig's DNA to remove the gene that tells it "I am a pig" so that genetically engineered and cloned pigs can be bred to use their body parts, such as liver, kidneys, and even the heart to replace failing organs in humans with little or no need for anti-rejection drugs! On Ripley's "Believe it or Not" program which was aired on May 25th, 2005, they were demonstrating new techniques for heart bypass surgery, where the doctor can do his work with a robotic machine and only three small incisions instead of the traditional open heart surgery that was much more risky. (But well worth the risk!) And new ground in medicine and medical science are happening faster than I can add them to this book, almost daily. And it all started in the last 100 years!

In the last chapter, we discussed computers and the effect they are having on everyday life. Every office has one, no matter how small, not to mention fax machines, e-mail, and copying equipment and on it goes. Banking is going more and more to a cashless financial system, and computer storage has gotten so massive with so small a space you can now hold a hard drive that will store 100+ Gigabytes in the palm of your hand. In 1968 there was a computer built in Belgium nicknamed "The Beast" that took up an entire floor of a building and could store all the information needed to identify every one of the (then) 2.5 billion people in the earth. Many thought this was the forerunner of the Anti-Christ. Today we have devices that you can set on your lap that will have that same capability.

Another amazing phenomenon is the worldwide population explosion. Dr. Harold Sala, in Guidelines for Living states,

"At the beginning of this century, the world's population was approximately 1.7 billion people; today, it stands at slightly over the six billion mark, thus making the explosive growth of humankind in a century one of the most remarkable phenomena of all times. Here's how you can put this in perspective:

In the first century, the world's population was 250 million people. It took 1500 years--from the time of Christ to the time of Martin Luther--for the population to double reaching the 500 million mark. Then in the next 300 years, which takes us to the year 1830, the population doubled, reaching the one billion mark. From that point on the increase in population soars dramatically.

Here are the statistics in brief. Between 1830 and 1930, the population doubled again, reaching the two billion mark. In 1960--only 30 years later-- another billion had been added to the world's population even at a time when millions died in World War II. Between 1960 and 1976 as birth control pills became widely used, the population still soared, reaching the four billion mark. Today as the twenty-first century begins, our population stands at over the six billion mark giving scientists one of the greatest challenges ever facing any generation: how do we feed and cloth such burgeoning masses of men and women, to say nothing of what we will need to do to prevent further pollution of our air and water?"[1]

Dr. Sala's article goes on to tell us that communications has also grown in almost direct proportion to the population explosion. At the start of the 20ᵗʰ century we communicated with one another by mail as discussed earlier in this chapter. Today that has now gone to paperless electronic communications, which includes radio, television, email, the Internet, and on it goes. Do you remember the old Dick Tracy comics? Dick used a 2-way wrist TV to talk to his headquarters and contacts. My last credit card statement included an offer for a hand held television with a newer brighter type LCD display that produced razor sharp pictures. How much longer do you think it will be before we actually have what Dick Tracy's author **imagined** only 30 years ago? I would expect to see it in just a few more months. I fully agree with Dr. Sala when he links the communications explosion to the population growth and points out that God is orchestrating the entire program. But what's really important is the time frame when the bulk of these events have taken place: All in the last 100 years!

Education has undergone some major changes. When I was a teenager, I looked forward with great anticipation to my high school graduation. Finally, that great day came (1964) I had arrived! Today, that same status applies only to the day you receive your college degree. In the 1940's and 1960's a person with a high school diploma was considered "educated". Today, a four-year college degree is the minimum requirement. Don't believe this? Then try getting a decent paying job without a college degree. It just will not happen in today's society. Perhaps it's the result of using the Public School System as a laboratory in which to conduct social experiments or maybe it's just that with the advent of the digital age, educational expectations are much higher on the part of the business community. Regardless of the reason, the end result is the same: "Men shall run to & fro about the face of the earth and knowledge shall be increased..."[2]. We are certainly seeing that happen today.

---

[1]. ___Guidelines for Living Radio Broadcast___ - Dr. Harold Sala Air Date: 1/5/00

[2]. ___Ibid___ – See Chapter 6, Note 3

Transportation only 100 years ago consisted primarily of horses and carriages. The "horseless carriage" was just on its way to mass production. Today we drive at speeds exceeding 70 MPH on Interstate highways. Passengers are shuttled across the country and around the world at speeds of 500 - 700 MPH in commercial airliners and the military has aircraft that can easily exceed that. As far as space travel, (the subject of science fiction 50 years ago) we have sent men to the moon and back and have sent robotics devices to other planets, even a little car that drove around on the surface of Mars and sent back astounding color pictures of the "Red Planet" Because of space exploration we now know that the rings of Saturn are made of an array of gases, the complexity of which still amazes the scientists that are studying them.

And what about inner space travel? For 84 years the remains of the great liner Titanic lay at the bottom of the ocean, undiscovered, unvisited by man, nearly 2 miles beneath the surface of the Atlantic Ocean. In the early 1980's a team of Navy engineers and private scientists began building submersible ships designed with cameras to photograph the floor of the ocean at depths where the human body would be reduced to something the size of a bucket of ash instantly. And yet there are fish and crabs swimming around down there! (Wonder how God did that?) In more recent years, (1994-Present) they have built ROV's (Robotic Operated Vehicles) that can probe deep into such wrecks and learn the secrets that the ocean has held to itself for centuries. All in the last 100 years!

We live in exciting times. I would not want to live at any other time in history. Yet we live in dangerous times. We have become slaves to the things we created. Recently, in the aftermath of Hurricane Isabel, I experienced a three-day power outage. You've all been there. What to do? Can't watch TV, listen to the radio, or even work on the computer. In fact the computer went down with the power and it took a significant recovery effort to restore it afterwards. So I cuddled up with a good book and read by candlelight. How crude. I had to tolerate 85-degree heat, with high humidity and no air-conditioning. How disgusting. How could they do this to me? See what I mean. What became an unpleasant reading experience for me was an enjoyable time of relaxation for my grandfather.

Please understand the point of emphasis of this chapter. Many of the things I noted were also mentioned in previous chapters, but the important thing I want you to see here is the time span. We have nearly 6000 years of recorded history to consider. These advances are not the product of half that time; not even one quarter of that time; not even one millennium. Yes, we have indeed come a long way, all in 100 years! We have seen a technological explosion in the last 100 years in which man has expanded his knowledge of the structure of the universe and the things around him more in the last 100 years than the 5900 that preceded it!

CHAPTER 10:
1000 YEARS

As we began the 21st Century there seemed to be a new fascination with the fact that we were beginning a new millennium. All kinds of predictions were being made and who knew what to expect? Computer designers and programmers had been preparing for the data storm of our lifetime since many programs and systems were set up with the old military style Year+Month+Day indexing scenario. When searching an index, a computer places the lower numbers first in the index database. That means that 00 (2000 AD) ends at the top of the index and 99 (1999) was at the bottom. For this reason a new phrase - "Year 2000 Compliant" had entered our vocabulary and had become a part of many contract and/or equipment purchase requirements. "Y2K" (Year 2000 Bug) became a household word and was showing up everywhere. Computer analysts made big money fixing it for banks, corporations, hospitals, and virtually everything else that is so dependent on computers. I spent hours scanning my databases and the thousands of lines of code for programs I had written to make sure that at 12:01 AM on Jan 1, 2000 my computer didn't crash and half my programs with it. Fortunately for all of us, the transition went pretty smooth with only minimal glitches to small systems. The flow of essential utilities and services was unimpeded by the entry into the new millennium.

But why was there so much pre-occupation with the year 2000? Is the "end of the world" as we know it on the horizon? Will the long anticipated return of Jesus Christ in His Millennial Kingdom become a reality? I think it would do us well to take a good look at the big picture here. If something of great spiritual significance, which will alter the course of the lives of every man, woman and child on Planet Earth, is about to happen, then you need to know all the details... who, when, where, & how.

A good idea in considering these facts is to link them together with a key verse of Scripture found in II Peter which states that "A thousand years is as one day with the Lord and one day is as a thousand years"[1] According to the Biblical record in Genesis, God created the world in seven days.[2] There have been many theories about the length of these days and it is not the purpose of this book to enter the "Creation vs. Evolution" arguments. In light of our key verse however, we should consider the important fact that time is something related more to man than to God. In eternity there is no reckoning of time. When you reach your eternal destiny, whether it is Heaven or Hell, you won't be wearing your Rolex™.

If, in fact, God considers a day as a thousand years, then lets look at time as we have it recorded in Scripture and history in that light. Since the creation of man in the Garden of Eden we have seen three major periods of time elapse, each containing two thousand years. From Genesis 1:1 to Genesis 6, two thousand years passed with little recorded history other than what we have in Scripture. Then around 2000 BC came the great universal flood in which only Noah and his family and the animals he took on the ark survived. Following this came another period of two thousand years to the birth of Christ. This event was so historically important that we now reckon time by it. This was no accident in God's plan for the ages, as we will soon see. The years before Christ are called BC (Before Christ) and those since are called AD. Since His Redemptive Mission to earth which changed men's lives forever, (and still does!) we have seen two thousand years pass. This total of approximately 6000 years is a span of six "days" by God's reckoning, Could we now be entering what could shortly become the seventh "day" of God's master plan? If so, what significance could the number seven possibly have?

Throughout the Bible the number six is often seen as the number of man. Take Revelation, for example where it states that the number of the Anti-Christ will be "666" and that is referred to as the number of a man. In 2000 AD we will have completed six millenniums. Is man's time about to be up? I think so. Another important number in God's plan is the number *seven*. We often think of seven as the Number of God or the number of perfection. In Chapter 2 we talked about the *Seven Churches* in Asia Minor and related that to the *seven periods of church history* which we have seen pass since The Day of Pentecost. There are also

---

[1] *II Peter 3:8-9* - But you must not forget, dear friends, that a day is like a thousand years to the Lord, and a thousand years is like a day. 9The Lord isn't really being slow about his promise to return, as some people think. No, he is being patient for your sake. He does not want anyone to perish, so he is giving more time for everyone to repent.

[2] *Genesis 1:5* - God called the light "day" and the darkness "night." Together these made up one day.

mentioned in Revelation "*Seven Golden Lampstands*"[3] and many of the judgments to come in the Seven Year Tribulation period (see Chapter 4) are going to be poured out in sevens. There will be a scroll opened which is sealed with "*Seven Seals*"[4] and "*Seven Vials of Wrath*"[5] will be poured out destroying nearly 2/3 of the people living on earth at that time.

So what am I leading up to? If six is the number of man, seven is the number of God, and God reckons time in blocks of 1000 years, could it be possible that we are in the final countdown as we enter the seventh millennium? Is man's time up? Is God's predicted Millennial Reign about to begin? And how close does that place us to the coming of Jesus Christ, and the Rapture of the Church? Read on... It gets better, or more frightening, depending on where you stand with God.

Many people today are entering the 21st century with fear. Others are beginning it with joyful anticipation. What about you? Will you be taken? Or will you be left? When is it going to happen? Will it be in our lifetime? Will it be another 1000 years? The next Chapter will help bring that into focus as we consider the parable of the fig tree and then the exciting summary of all our findings in the following chapter where we literally "put it altogether".

[3.] **Revelation 1:20** - This is the meaning of the seven stars you saw in my right hand and the seven gold lamp stands: The seven stars are the angels of the seven churches, and the seven lamp stands are the seven churches.

[4.] **Revelation 6:1** - As I watched, the Lamb broke the first of the seven seals on the scroll.

[5.] **Revelation 16:1**- Then I heard a mighty voice shouting from the Temple to the seven angels, "Now go your ways and empty out the seven bowls of God's wrath on the earth."

CHAPTER 11:
UNSEALING THE PROPHECY

A proper examination of Bible Prophecy demands an understanding of several factors. These considerations will help to evaluate the prophecies of God's Word and what they mean to us and should be applied to each individual prophecy we are going to study. They are:

1. Who wrote the particular prophecy we are examining? Was it written or given by an Old Testament prophet, a New Testament apostle, or by Jesus Christ Himself?

2. When, where, and under what circumstances was it written? Was it written as the result of a vision or as instruction to a particular group of people, such as in the Olivet Discourse given in Matthew 24 & 25 by Jesus?

3. Was there any specific purpose for which it was written? For example, consider our discussion of the "Seventy Weeks" prophecy in Daniel Chapter 9, where a specific set of reasons (already listed and discussed at the beginning of chapter 5) is stated. Here, Daniel gives a specific set of things that would happen when the time came to fulfill this prophecy.

4. Were there any specific instructions related to the prophecy given by or to the prophet as to when the prophecy would be opened to our understanding, such as in Daniel 12:4? It is precisely this subject that I would like to expand on in this chapter.

In light of the above statements, I would like to direct your attention to two specific instances where an instruction is given concerning a particular prophecy or group of prophecies. They are found in Daniel 12:4 and Revelation 22:10. The Daniel passage says,

"But thou, O Daniel, shut up the words, and seal the book, even to the time of the end: many shall run to and fro, and knowledge shall be increased" *(Daniel 12:4,King James Version)*

Now let's note the Revelation passage where it says,

"Then saith he (the angel) unto me, Seal not the sayings of the prophecy of this book: for the time is at hand" *(Revelation 22:10, King James Version)*

Oops! Have we stumbled across a contradiction? Of course not: There are no contradictions in the Bible. It is God's inspired, inerrant, and final Word to mankind to prepare the world for the Church Age or "Day of Grace" as we discussed in Chapter 8 where we dealt with the "Times of the Gentiles" as spoken of by Christ in the Olivet discourse. And therein lies the key to understanding why one prophet is told to seal the prophecy and the other one is told to leave it unsealed. If we use our little guideline above, we will find that an Old Testament prophet wrote the prophecy in Daniel at least 500 years before Christ. A New Testament pastor, the apostle John, while in exile on the island of Patmos, wrote the prophecy in Revelation. And this helps us to understand why this is not a contradiction.

The first was written before Christ. The Olivet discourse, given by Christ, which we will examine in detail shortly, had not taken place at the time of Daniel's prophecy. Neither had the redemptive work of Christ in the death, burial, and resurrection occurred yet. But in the Revelation passage these great events were now history. In other words, Daniel was looking at things to come in a totally different light than John. Remember now, both of these prophecies were given by direct revelation, (plenary verbal inspiration is the theological word for it) so, in spite of the differences in time there has to be a reason for these differing commands regarding the sealing of these prophecies.

We must, therefore, find an answer other than the time or person involved in the penmanship of these statements. Timing is important, however. The prophecy in Daniel was written to us, the people of the 21st Century, because some things had to happen ***during our lifetime*** that would cause the prophecy to become unsealed. For any generation prior to ours, an understanding of the meaning of the prophecies of Daniel 9 and 12 would be nearly impossible. The time had to be right and as stated in Daniel 12:4, it would have to be the time of the end. Is it that time now? To determine that, we must quickly examine the two things in that verse that delineate exactly when that time would be.

First of all, it says that men would run to and fro. Today, a businessman can eat breakfast in New York, lunch in Paris, and dinner in Jerusalem. Would you consider that as a fulfillment of one of the two conditions? I think so. There would be no other plausible explanation of that part of the verse. Now let's consider what brought our hypothetical businessman on this seemingly wild adventure to start with. We look in his briefcase and find out that he is a nuclear physicist who is being sent to Israel to help train their doctors in Nuclear Medicine Technology so they can administer the same treatment to Jewish heart patients that I received in the Nuclear Medicine Lab at the local hospital where they injected a radioactive dye into my bloodstream and watched it move through my heart and its associated arteries to check for blockages. Would you say that the second half of that requirement (knowledge shall be increased) to unseal Daniel's prophecies had been met? I certainly would.

Do you see what I am getting at here? An examination of Daniel's prophecies of the world governmental structure and knowledge explosion would have been meaningless to any generation before ours. God therefore, in His infinite wisdom chose that those prophecies remain sealed until the time of the end or better known as the "last days". God not only instructed Daniel to seal them, but He also told him when they would be unsealed. And now that the time of the end is near, they have been unsealed and we have the fulfillment of the two conditions attached to them that would prove that they are, in fact, unsealed. In chapters 6, 7, & 9 we dealt with the knowledge explosion, the point of sale system we now have where one person can control every transaction you make and choose to prevent you from being able to buy or sell, and how the identification chip would be a part of that system and would be implanted in the hand, forearm, or forehead. And by the way, in a recent newspaper article, it was stated that they were now using these implanted chips on people. That's right; they are now planning to use it in Alzheimer's patients so that if they wander away from their home or nursing facility, authorities can scan them and see who they are and where they belong.

> "The Food and Drug Administration approved the Veri-Chip for use in people last October. An applied Digital Solutions spokesman estimates that about 1,000 people have already had a Veri-Chip implanted, usually in the right triceps. At the moment, it doesn't carry much information, *just an identification number* that health care professionals can use to tap into a patient's medical history." [1]

Remember our discussion in Chapter 7 on how a simple hexadecimal number could be used to identify every man woman & child in the world? *Wow!* How close are we really getting?

The command to John in Revelation 22:10 was far different. The book of Revelation started out by describing seven churches in Asia Minor, all in existence at the time of the writer's exile on the island of Patmos. The

---

[1]—*The Virginian Pilot* [*Commentary Section J1*] "Take my privacy, please!" *by Ted Koppel* 6/19/2005

remainder of the book would delineate many of the events that would occur during the Tribulation Period, already discussed in earlier chapters of this book. It had to be left unsealed so that the conditions and progression of the Church during the "Day of Grace" could be understood and expounded by Bible teachers worldwide. The information given about the things that would happen during the Tribulation period were left unsealed so we could see that we are still in the "Day of Grace" and have a few more precious years to share the Gospel. And last but not least, Revelation contains the only detailed account of Satan's final defeat and eternal punishment in the Lake of Fire. This is information that helps us on a daily basis to know and understand what the final outcome of Satan's short reign on earth would be and that he is a defeated enemy. So God, in His infinite wisdom chose to leave the prophecies of Revelation unsealed.

In summary, the prophecies of Daniel were written by an Old Testament prophet about people of the 21st century and would likely be understood only by those who now live in the time of the end where the unsealing of those prophecies has taken place. Revelation, on the other hand, was written by a 1st Century prophet about what would most likely happen in the 21st century, but was written to people of all centuries following its writing to encourage them to carry out God's mission to reach a lost world with the precious Gospel message. We *are* that final generation. And in the next couple of chapters you will understand why I believe that.

CHAPTER 12:
THE FIG TREE

" Now learn a parable of the fig tree; When his branch is yet tender, and putteth forth leaves, ye know that summer is nigh: So likewise ye, when ye see all these things, know that it is near, even at the doors. Verily I say unto you, this generation shall not pass, till all these things be fulfilled." Matthew 24,32-34

In order to understand the meaning of this statement, we must analyze what it says, when it was spoken, to whom it was spoken, and the context in which it was spoken. The passage of Scripture here was part of what is often referred to as "The Olivet Discourse" That section starts in Matthew 24:1 and ends with Matthew 25:36. It was directed toward the nation of Israel, and can be summarized in eight parts. In the first section, Christ tells His disciples that Jerusalem would be destroyed. The devastation would be so great that Christ said, "There shall not be left here one stone upon another, that shall not be thrown down."[1] We now know that to be a historical event of the past, which occurred in 70 AD when the Romans destroyed Jerusalem. In their search for gold and other treasures, some of which was hidden in the walls of the Temple, the Romans totally disassembled every building in the Temple complex down to the ground, fulfilling Christ's prophecy **_exactly_**. The next section presents the question by His disciples as to when it will happen. Following that is a discussion of the conditions the world would see during the Church Age, which began with the Day of Pentecost and still exists today.

Look at some of the things He pointed out here: "Many shall come and say I am Christ; and shall deceive many.[2]" Remember Jim Jones? And

---

[1] **_Matthew 24:2_** - But he told them, "Do you see all these buildings? I assure you, they will be so completely demolished that not one stone will be left on top of another!"

[2] **_Matthew 24:5_** - For many will come in my name, saying, `I am the Messiah.' They will lead many astray.

what about many of the other cults that have led hundreds away? "And ye shall hear of wars and rumors of wars... For nation shall rise against nation and kingdom against kingdom:"[3] Sound familiar? Remember our discussion in Chapter One about wars and the "War to end all wars" and how many there have been since? (There has not been a day in which a war is not going on somewhere since I have been born.) "There shall be famines and pestilences."[4] What about the graphic TV coverage of the starving children in Ethiopia and Africa, and even still today those infomercials for the many organizations trying to help a starvation problem that seems to have no end? Yes, with the worldwide population explosion there are already almost more people on this earth that there is food to feed them; and the situation is not going to improve as more ways are being found to help people live longer. "Earthquakes in divers places..."[5] In the last century there have been more earthquakes recorded than in the entire span of time prior. Is it because of the advent of high tech recording equipment? Or could it be that there has actually been an increase in seismic activity since 1900? I think the latter is the case because the resulting death and devastation from such huge earthquakes as we have seen this century would have been recorded in historical documents with or without the geological breakthroughs of the last few years. Just a few months before I finished this book an earthquake in Turkey claimed 18,000 lives and left countless injured and homeless [6]. Then, about three weeks later, they were hit with another one in the Northern part of the country. The devastation was nearly unbelievable. And what about the earthquake driven Tsunami of 2005 that saw a death toll approach one million men, women, and children? And the worst is yet to come.[7]

There are some other things that must occur during this period of time in which we live. "Iniquity shall abound." Remember Chapter One of this book? "The love of many shall wax cold" See Chapter Two. "And this Gospel of the Kingdom shall be preached in all the world for a witness unto all nations; then shall the end come." Today we have the means to do exactly that! With a world wide satellite network in place, and with an abundance of Christian radio and television stations, networks, and programs it is now possible for every man, woman, and child to hear the message of Salvation. The next event on God's calendar: "The end (of the Age of Grace) will come"

[3.] ***Matthew 24:6*** - And wars will break out near and far, but don't panic. Yes, these things must come, but the end won't follow immediately.

[4.] ***Matthew 24:7*** - The nations and kingdoms will proclaim war against each other, and there will be famines and earthquakes in many parts of the world.

[5.] ***Matthew 24:7*** – *Ibid Note 4 Above*

[6.] ***NBC Nightly News*** Air Date: 8/30/99 6:30 PM EDST

[7.] ***Matthew 24:8*** - But all this will be only the beginning of the horrors to come

The fourth section of the Olivet Discourse addresses the seven year period that Bible scholars have come to know as the "Great Tribulation" in which there will be peace on earth for 3 1/2 years followed by the revelation of the Anti Christ and "great tribulation, such as was not since the beginning of the world to this time, no, nor ever shall be." Because we have already discussed this time period in previous chapters I will not be redundant in adding more here, save to say that it is a time when the Nation of Israel will suffer greater hardship than ever before, even during the Holocaust under Hitler!

The next event in the chronological order of events in this passage details the Second Coming of Christ to bring an end to the Tribulation period, the defeat of Satan and his Anti-Christ, and the beginning of Christ's reign on earth. You can almost feel the tension building here as Christ unfolds these future events to His disciples, so it is here that he now answers the question they asked Him in the second part of this discourse, "Tell us, when shall these things be? And what shall be the sign of thy coming, *and of the end of the world*?"

In order to understand the meaning of the parable of the fig tree, you must understand, as we said that this entire text of the Olivet Discourse is directed toward the Nation of Israel, a nation that has, since the destruction of Jerusalem by Rome in 70 AD, been scattered throughout the world *even through the Nazi Holocaust!* Could the Holocaust have been the Great Tribulation Jesus spoke of? I don't think so. In spite of the fact that six million Jews died under the evil hand of Hitler, the time was too soon. There was not even a "Nation of Israel" in place during that event; which makes the Great Tribulation an event yet to be played out in the future.

In 1948, two years after I was born, the flag of Israel was hoisted for the first time over the land of Palestine since 70 AD. The fig tree now has leaves, the branch is tender, and summer is near. The re-establishment of the nation of Israel in their homeland is the pivotal event to the future: YOUR FUTURE! (And mine). This is an exact fulfillment of several Bible prophecies, some of which we have discussed in earlier chapters. It is my firm conviction that this is the event Christ spoke of in His parenthetical pause during the Olivet Discourse, where he used a fig tree as an example of a major change of the season. But there's more. Christ then goes on with sections six, seven, and eight of His discourse telling of the Judgment of the Nations, especially Israel and the non-believing nations of the world. (Remember the true believers will have already been removed prior to the Great Tribulation) The "Times of the Gentiles" are finally nearing an end. The "Valley of dry bones" of Ezekiel Chapter 37 has come to life!

CHAPTER 13:
LET'S CRUNCH SOME NUMBERS

Perhaps the most significant part of the statement about the fig tree is the time frame to follow: "Verily, I say unto you that this generation shall not pass, 'till all these things be fulfilled." Matt 25:34. So now comes the big question: When does this generation start, how long is it, and when does it end? If we can answer these questions, we can with reasonable accuracy, estimate when it will happen. Please understand from the beginning of this discussion, I am not a date setter. If you expect me to tell you (as some have) which day, month, and year Christ will return, then you can stop reading here and throw this book away. My whole purpose is to bring everything we have discussed in this text together to show you why I think it is going to be soon, very soon, most likely *in my lifetime* and maybe yours, if you are alive and walking around on Planet Earth today.

First of all, we should carefully consider the condition of the world as discussed in chapter two and the first three parts of the Olivet Discourse. Then we must consider the condition of the Church as discussed in chapter three and part three of the Olivet Discourse and many other places in Scripture, too numerous to detail here. Then we need to remember the prophetical time lines detailed in Chapters 4 and 5. We need then to consider how the stage is currently being set in our lifetime as detailed in Chapter 6. And finally we need to see how much things have changed in the last 100 years as opposed to all the previous millennia combined. Certainly, the fact that the technological explosion detailed in Chapter 6 and predicted in Daniel 12:4 has happened *in our lifetime* is no accident.

Then we must consider the arrangement of the Millennia as part of the big picture. Is it an accident that 6,000 years have passed since man was placed in the Garden of Eden; and that the number of man in Scripture is 6 or the number of the Anti-Christ is 666 in Revelation? If the number of God throughout the Bible is seven and we are beginning the seventh millennium, could this also be a coincidence? But as mentioned earlier, the pivotal part of our discussion is the Olivet Discourse and the verse that states that all these things will come to pass in our generation; that is, the generation in which Israel became a nation; the generation is which the fig tree "putteth forth it's leaves". Any one of these things alone would be enough to generate excitement in the heart of a Christian or fear in the heart of an unsaved person, but when they all are taken together, it really starts to look like something is about to happen.

When I was a child, one of my favorite toys was an erector set. It consisted of strips of steel with holes, nuts and screws, corner pieces, baseplates, winches, etc. I would first build a boom. Then I would put some more pieces together and build the column section. After this I would assemble the pulleys and winches. Then came the base plate and supporting sections. Then finally I would put all the sections together and behold I had a beautiful crane with which I could now build miniature skyscrapers that existed in my childish imagination. I now had the machinery to turn my dreams into reality, on a small scale of course! In the same way, God has given us the sections of His Plan for the Ages, scattered throughout Scripture, some occurring ages ago, some occurring in the last 2000 years and many more happening in the last 100 years with an almost unbelievable number *occurring in our lifetime!*

So now, we have already established that the generation Christ was speaking of began in 1948. So how long is it? Well, according to the dictionary, it's about 30 years, or the time that lapses between the death of one person and the birth of their son or daughters child. That would mean that the Great Tribulation should have started around 1971 and Christ should have come around 1978. Obviously we can't use the dictionary's definition of a generation. A better source would be the Bible. Always remember that the Bible is the best commentary on the Bible. According to Psalm 90:10 a generation (or average lifetime) is seventy or eighty years. I personally like 70 because that is closer to today's averages. (73 for men and 76 for women)

If you add 70 to 1948 you get 2018. Very interesting. Now take the year 2000 from 2018 and note that the difference is 18. Now deduct 18 from 2000 and note the result: 1982. What happened in 1982 that is so rare that it happens only once in every 176 years? We had a planetary alignment in our solar system. That's right... all the planets in our solar system, the only one in which God chose to place man, were in a straight line. Not only did one occur in 1982, but also it happened again on May 5th, 2000, except that in this one all the planets were behind the sun when looking at the line from earth. There have been two planetary alignments in 20 years and probably will not be not another one for nearly 200 years, if ever again. Could this be the start of the final countdown? Could God be telling us this is it? If so, then all the things predicted in Matthew 24 *including the Great Tribulation* will come to pass soon. Very soon, possibly by the year 2018 or 2028.... *in my lifetime!*

Does this mean that the Rapture will come seven years prior to that or sometime before 2021? Possibly. It could come today. Or it could be several decades. Only God knows for sure. Remember, determining the size of a generation is not an exact science. The Scripture passage we quoted earlier in this chapter, (from Psalm 90:10), states that man's days shall be threescore and ten, and, if by reason of good health they could be fourscore. So, as you look at all these numbers, you may be thinking, "There he goes trying to set a date." **ABSOLUTELY NOT!** I am just trying to show that I believe that these events are imminent, just around the corner... the YEAR 2000 corner! We're here folks. All the major pieces of the puzzle are now in place and what a beautiful picture we now have. Just think of the possibilities. You may very well be a part of the generation that will not have to die. If you know the Lord and have a personal relationship with Jesus Christ please don't waste your money on a cemetery plot, at least not right now. PERHAPS TODAY the trump will sound, maybe even before you finish reading this book.

CHAPTER 14:
WHAT IF... AND "WHAT ABOUT ME?"

Some of the most powerful words we have in the English language (and most other languages that are structured similarly) are the shortest. Take the word "if" for example and look at some of the many statements you have heard:

" If I just had gotten here sooner...."
" If it just didn't cost so much..."
" If I just didn't feel so bad............"

"If" when combined with "and" can often denote a combination of events which must occur for something to take place:

"If I had just gotten here sooner and it didn't cost so much and I just felt a little better..."

I've often said, "If cats had wings they could fly like eagles and wouldn't have to walk on fences." Well, that may be an exaggeration but it makes the point that a lot of things would be different *if* the circumstances that generated them were different. We often add words to our "ifs" to denote a singularity of events such as *"if only"*:

"If only I had just gotten here sooner..."* would indicate that of all the other possibilities this one event would have been the turning point. The year after I was born, my sister Jeannie, age 8, died of Polio, a crippling killer that was especially deadly to young children. Three years later Dr. Jonas Salk discovered a cure for that horrible disease and today it is almost unheard of. *What if* Dr. Salk had found the cure three years earlier? *What if* Jeannie, like myself, had been born eight years later? See what I mean? "If" is truly a small word with a big meaning. "If" can change the course of

history. *What if* Adolf Hitler had never lived or been elevated by his people to where he got the opportunity to do the things he did?

What if these things we discussed in the preceding chapters are true? Let's look at both possibilities: First of all let's assume that they are not and that nothing new is going to happen. Where does that leave you? Well, it would not make any difference how old you are, how old you would live to be or how you should live. If you are not expecting Christ to return and these prophetic events to take place, then life would be no different for you thirty years from now than it is today unless you die during that time and have to face your Creator and Judge. But what if these numbers have some meaning and we are, in fact, at the door, then where does that leave you? Well, that depends on where you're headed now. Heaven or Hell, the decision is yours.

The ***decision.*** And that brings up another interesting fact that can be linked to "If". The world runs on decisions. Every day of our lives we face those "ifs". We then must decide which one we will choose. Those decisions can change our lives, the lives of those around us, and even the history of a nation or the world. God loves you so much that He will never force you to do something you don't decide to do. Your response will determine your eternal destiny. And with the events we have discussed throughout this book unfolding at such a rapid pace, there may not be much time to waste. Recently I was in a checkout line discussing with another customer the benefits of using cash instead of checks while we waited for a lady to write her check, get it scanned, fill out her register, etc. He said, "Won't it be great when they get the little chip in your wrist that you can just scan at the register and checkout in seconds?" My friend, people are ready for the things we discussed in this book. Many are already beginning to unfold. Do you remember Jesus words in that Olivet Discourse we discussed? "And when these things ***begin*** to come to pass, then look up, and lift up your heads; for your redemption draweth nigh." (Luke 21:28)

You are one heartbeat away from eternity. You may live to see these events continue to unfold or you may not live to read the last sentence of this book. You have little control over that, but what you do have control over is where you will spend eternity. According to the Bible, you, like me, and every other person that has lived on this earth (except Jesus Christ) has sinned. (Romans 3:23) And because you have sinned, you must die; (Romans 6:23) not once but twice. The first death is the death of your body. The second death is eternal damnation in the Lake of Fire (Hell). Read Revelation 20:11-15. But God loves you. (See John 3:16 & Romans 5:8)

Now **you** must decide. Remember what we said earlier? How a single decision can change the course of your life? Well that is what these next few verses are all about:

God has a plan for your life. You can choose to accept His love or reject it and go on without Him. Now look up Romans 10:9,10. "If you will confess with your mouth Jesus as Lord and believe that God has raised Him from the dead, you **will be** saved. For with the heart man believes unto righteousness, and with the mouth confession is made unto salvation." Your church membership, good works, and best intentions won't do it. See Ephesians 2:8,9. So there you have it: God's simple plan of salvation. Simple faith in what Christ has already done. Study the Scriptures: He will do the rest.

If you have made this decision to accept Christ as your Savior you should immediately seek a Bible teaching Church that believes what you have read here and begin to study God's Word and find fellowship with those who teach these truths. If you have accepted Christ as your Personal Savior then you have something to look forward to as you read and re-read this book and see God's plan for the end times unfold right before your eyes. You should also begin a discipleship program to help you learn to study the Bible [1].

And if you haven't made that decision yet, then you should be shaking in your boots. We're **here** folks. This is it! The very next thing on God's prophetic calendar is the Sound of the Trumpet. Are *you* ready?

---

[1] Here are two links for organizations that offer excellent discipleship programs:

http://www.discipleshiplibrary.com/index.php (The Navigators Discipleship Page)

http://www.bgea.org/SpiritualHelp_Index.asp (Billy Graham Evangelistic Organization)

BIBLIOGRAPHY

All References from the Bible are taken from the New Living Translation unless stated otherwise.

The New Living Translation is published by Tyndale House Publishers © 2003 and can be found at their web address: www.newlivingtranslation. com *(Used by Permission)*

Chapter 3, Note 11:
Dr. Mike Windsor, CBI Studies, 4/18/05 Colonial Bible Institute, Va. Beach, Va. (757) 479-3706 *(Used by Permission)*

Chapter 9, Note 1:
Guiidelines for Living - Dr. Harold Sala 1/5/00 *(Used by Permission)*

www.ingramcontent.com/pod-product-compliance
Lightning Source LLC
Chambersburg PA
CBHW021240280526
45784CB00005B/2180